SYSTEMS OF NOURISHMENT

OPTIMIZE YOUR DIET

LUKE COMER

PREFACE

You might ask: Why should I read this book?

One word: your body.

You are stuck with it for the rest of your life. And that body, along with its attached brain, can either help you dance through life or torture you, literally, to death.

And in reality, nothing affects your body more than your diet over your lifetime, especially in combination with your lifestyle.

This book is about using diet to help you navigate your life, free from disease and pain, while feeling good and strong.

You might also ask: Why should I read this book and not others on the subject?

Because this book takes another and better approach. While other books encourage you to eat specific foods or nutrients, based on some limited understanding of this or that, this book teaches you to understand your diet comprehensively and then use that understanding to design your own diet.

In this book, you do not encounter dietary ideologies like Veganism, Vegetarianism, Paleo, Mediterranean, etc. In many senses, none of that matters to your body. Instead, you will learn how diet works inside your body, how diet worked in your evolution, and how diet is understood through the authorities and scientists running studies and making recommendations.

Is this too much for you to understand?

Not at all if you are committed to learning something so vital to your well-being. I did the long, arduous work, of reducing the cryptic complexity and technical jargon of science into grokable narratives and simple language that you can commit to memory. To riff on the old adage: I am not giving you fish but teaching you to fish.

I started studying diet about thirty years ago because of my own ailing and banged-up body and brain. Even though for much of that time I was not getting everything right, I still radically improved my own health and well-being, and now, towards the end of that journey, I am getting most everything right. In my own diet, I merely eschew most modern foods, sugary stuff, refined stuff, and fake stuff. Then I just eat natural foods to satiation, while most importantly, carefully selecting, preparing, and combining those foods in certain ways. For protein, my daily diet consists of dairy, eggs, small amounts of meat and seafood and lots of broth and other collagenous tissues—and small doses of liver. For carbs, I eat sourdough bread and pancakes made from whole barley, spelt, bean and nut flours, as well as my favorite: fresh fruit and berries. For fats, I eat combinations of animal fats, olive oil and various nuts while eschewing most plant oils. I cook stews several times per week, loaded with all the foods above, as well as vegetables and herbs, and consider them the most nourishing of all meals. I rarely use supplements.

I never curtail calories unless I am just not hungry, and never gain weight. I sleep well almost every night without waking. I do not have any health issues, dental problems, or chemical addictions, outside of some muscular and joint issues due to my lifetime of pursuing various sports like rock climbing, kitesurfing and tennis. I rarely get sick with any infections, maybe on average about once every four years or so. And I confess, even though this sounds strange, I have only visited the dentist once in several decades and the doctor maybe three times and never take prescription or over-the-counter drugs. And on top of that, I enjoy, in select sorts of ways, various psychoactive substances in moderation, including alcohol.

Believe it or not, as you shall learn, this is the biological norm for humans, but obviously not our cultural norm. This is the way I prefer to eat but you do not need to eat these same foods; as mentioned, you just need to understand your underlying "Systems of Nourishment" and then apply that to your American, Japanese, Mediterranean, French or even Vegetarian or Vegan diet.

Over all those decades of studying diet, I spent over half my time just dispelling the enormous amount of confusion that plagues the field from limited understanding, zealous ideology, bad science, special interests, and marketing jargon. So in this book, I save you decades of confusion, and possibly disease and malaise, by helping you dispel, much faster than myself, all this mendacity to understand the truth of your diet. I then present that truth within one system. If you understand that system and implement the resultant principles, all evidence suggests that you will develop a much stronger body and brain, while resisting not only the "diseases of civilization" that afflict most Americans but also many infectious diseases and countless other maladies as well. Beyond that, you can then further develop your biological potential and hopefully use that potential to better yourself and your world.

TABLE OF CONTENTS

INTRODUCTION

In this book, I use "systems science" to help you understand your diet and then apply that understanding to enhance the quality of your life. Generally, systems science is defined as using various forms of methodology to comprehensively understand both simple and complex systems in nature and society—and then using that understanding for some purpose.[1] In this case, the system is your diet and the purpose is you, the reader, your life, your health and your well-being.

More specifically, in this book, I study diet using four methodologies that are all grounded in reality, observation and science. These four methodologies are: nutritional physiology, ancestral consumption, establishment recommendations, and scientific studies. I then use these four methodologies to analyze the seven premises of the optimum diet, described below.

You can then apply these seven premises to your own diet to optimize your metabolism —that is, your capacity for energy as well as the maintenance and growth of your tissues. However, you can also apply these principals to other and additional purposes, including weight-loss, anti-aging, disease reversal, as well as enhancements in mental and physical performance for various endeavors. Once you understand the systems of your diet, you can then alter that system for various results.

THE METHODOLOGIES

Nutritional physiology

If you want to optimize your diet, you probably want to know how diet works mechanistically, right? You cannot build great computers or fix cars unless you know how they work. In the same way, you cannot optimize your diet without understanding how that diet works in your body. So I reveal the origins and form of nutrients in our food; and then what happens when we process, cook and digest our food into nutrients. I then reveal how those nutrients are transported in our blood, absorbed by our cells and then metabolized for various purposes-or in some cases, recycled, stored or converted into other nutrients. You will understand the journey of your food from beginning to end.

Ancestral consumption

If you want to define your optimum diet, you probably also want to know what your ancestors ate, especially if you understand the importance of evolution in diet. By natural law—that is, through "natural selection"—all animals, including humans, evolved to consume foods that help them survive, thrive and reproduce. Evolution does not design animals to eat foods that make them sick, tired, and stupid. For example, fruit-eating monkeys develop traits that help them acquire, digest, and metabolize fruit, which then provide them with advantages in their survival and reproduction. For this reason, you should understand the foods you evolved to eat because those foods optimize your health, and other foods probably do not.

I first explain the likely diet of what I call "our theoretical common ancestor" (aka Chimpanzee-human last ancestor) that launched our line of evolution towards humans.[2] This hominoid existed around seven million years ago in the jungles of Africa, eating diets that were likely similar to the modern chimpanzee, or more specifically, the savannah chimp. I then trace how our diet evolved from there through two key hominoids, Australopithecus and Homo erectus, and then all the way to Homo sapiens through three different eras: foragers (aka hunter-gatherers), agriculturalists, and ourselves, moderns. Through this story, you see how we all evolved towards consuming certain foods and nutrients prepared in certain ways and in similar ratios to each other. You also see the overall pattern in our diet as well as the considerable variations in that diet—and then how to explain that variation.

Establishment Recommendations

If you want to design your optimum diet, you probably want to consider the advice of the experts—that is, the "nutritional establishment," consisting of the scientific and governmental authorities who are experts in their field. These experts do not define the optimum diet, but rather the recommended diet to prevent malnutrition and disease. They also sometimes determine adequate as well as tolerable intakes at the upper level for certain nutrients. Despite our differences in goals, I nonetheless strive to learn from them as much as possible.

Various governmental and scientific organizations participate in these determinations, including The Food and Nutrition Board of America, the National Academy of the Sciences, the European Food Safety Authority, and the World Health Organization. Their various experts compile enormous amounts of information from various scientific studies, and then draw conclusions on those studies and make their recommendations, and, interestingly, all these various organizations tend to agree with each other. So, for that reason, we should understand and consider their advice while also maintaining some skepticism.

Scientific Studies

If you want to understand your optimum diet, you probably also want to understand the plethora of scientific studies of many sorts, usually published in reputable and peer-reviewed journals. Of course, the "establishment" uses many of these same studies as sources for their recommendations. However, in many cases, they seem to disregard certain types of studies, especially ones that are more recent and, in that case, I tend to focus more on them. These studies range in their design and include many different types, including observational studies, epidemiological studies, intervention studies, genetic studies, clinical trials, and meta-analysis. While in many different forms, most of these studies operate in the same way—that is, they are attempting to trace the effect of nutrients, usually just one, isolated nutrient, on the health or disease of groups of people over some period of time.

Methodologies summary

While all these methodologies are excellent ways to understand the truth about our diet, they nonetheless all have their limitations. Regarding nutritional physiology, scientists understand, incredibly well, how nutrients function inside our bodies; however, they do not yet know much about nutrigenetics—the variation in the function of these nutrients from one person to the next, as individuals and various groups. Regarding ancestral consumption, scientists are working to reconstruct the diets of many hominoids and humans that do not exist anymore, so they have to reconstruct their diets based on limited and indirect evidence—and thus incomplete information. Regarding establishment recommendations, despite their plethora of experts, these organizations are generally subject to all sorts of mistakes, many now already proven, related to bad science, political and financial pressure and the generalized conformity and dogma that afflicts these organizations. And scientific studies are subject to multiple problems: faulty designs, bad conclusions and many, if not most, of these studies are funded by organizations with interests in specific outcomes. All of this explains in part why the study of diet is so bewildering. But in my case, because I am cross-referencing one methodology to another, I am more likely to identify the problems in one of the methodologies and find ways to correct those problems to reveal the larger truth.

THE PREMISES

As noted before, I will use these four methodologies to describe and analyze the seven premises used to design the optimum diet. Below I will explain each of these premises in brief and then, of course, develop them in greater detail in the remainder of the book.

Premise 1: Do no harm

Premise one, borrowed from the Hippocratic Oath, is: "Do no harm." In designing your diet, you should exercise caution in putting compounds, including foods, supplements, pharmaceuticals and the like, into your body that cause harm in one way or another, even when some possibly causing some good. Generally, the rule here is that, if your ancestors did not consume this substance, you probably have not evolved to utilize that substance, and thus you should exercise caution.

Premise 2: Optimize digestion

Premise two is that you should design your diet to optimize your digestion—that is, to make your digestion both complete and efficient. By "complete," I mean foods that are digested as completely as possible into individual nutrients and then absorbed into your bloodstream, which prevents these nutrients from flowing into your colon and producing various toxins and detriments. By "efficient," I mean foods that reduce your "digestive metabolism" (or "diet-induced thermogenesis")—the energy used by various organs to digest our foods.[2] When this happens, you use less of the energy in your food for digestion and more for other forms of metabolism, which, overall, then provides you with more energy for your life. However, we moderns now reduce digestive metabolism by removing the pulp, fiber and germs from our food but this also removes many nutrients from our food—which then causes us detriments. So you should instead improve digestion through other forms of processing that actually preserves and even increases the amount of nutrients.

Premise 3: Reduce antinutrients and toxins

Premise three is that this diet reduces antinutrients and toxins in any plant foods. Plants are just like us: they do not want to be eaten, especially into extinction, so they evolved ways of protecting themselves from their predators, including humans, thus discouraging us from eating them. Antinutrients interfere with our digestion and absorption of nutrients contained within plants.[3] Toxins outright poison various tissues of our body like our organs, brain, and skin.[4] Wild plants are especially rich in these compounds but even our domesticated plants still contain plenty of them. But through various techniques, you can reduce these compounds in your food so that you are affected by them minimally.

Many plants additionally contain storage proteins, including gluten and many others, which only serve to store amino acids for the plant's later use. However, it does seem that plants also evolved these proteins to make them extra hard to digest, so they also act like antinutrients—though not technically defined that way. In any case, you should design your diet to minimize these proteins as well.[5]

Premise 4: Consume all nutrients

Premise four is that this diet contains all nutrients needed to optimize all metabolisms, including all macronutrients, micronutrients, minerals, as well as fibers, and phytonutrients. While the types of foods we consume are important, they in some senses do not really matter all that much because, at the end of digestion, these foods are reduced entirely to their nutrients. For that reason, you should design your diet to contain foods that, in turn, contain all the nutrients you need.

We tend to sometimes demonize one nutrient and celebrate others—even somewhat randomly. Sometimes, sugar is evil; sometimes fat, and then especially saturated fat; sometimes omega-6 is bad, and omega-3 is good and then reversed back again. We tend to think of nutrients sometimes like our sports or movies: full of winners and losers, heroes and villains. But in the end, there is no distinction between good and bad nutrients—they all serve their purpose in the proper amounts and ratios.

Premise 5: Optimize calories

Premise five is that this diet encourages you to eat all the calories you need to optimize your metabolism—not more, not less. When eating too many calories, we tend to gain weight while trending towards the cluster of symptoms called "metabolic disease," and "diseases of civilization" that includes overweight, hypertension, cardiovascular disease, type 2 diabetes, autoimmune diseases, tooth decay, and some cancers. If you have these diseases, to one degree or another, you probably are more susceptible to many other maladies as well, including some forms of mental illness, like depression and anxiety, as well as infectious diseases.[6] But we tend to overeat because we are eating the wrong foods, in the wrong ways, with imbalanced nutrients. Because our body is not getting the needed nutrients, we just keep eating. When eating too few calories, we then slow our metabolism, which, over time, possibly leads us towards other complications such as under-nourishment. So this diet encourages you to consume the right number of calories to optimize your metabolism and your potential for life.

Premise 6: Include all nutrients in their optimum amounts and ratios

Premise six is that this diet contains all nutrients in specific amounts and ratios to each other that correlate approximately to the way our bodies utilize those nutrients — which in turn reduces waste and inefficiency while optimizing your metabolism. Due to the complexity and girth of premise six, more than half of this book is devoted to this subject.

As noted in the section above, we humans need all nutrients — so none of them are inherently or individually bad or good. However, they can become bad or good depending on the amount consumed and their ratios to other nutrients. Ordinarily carbs and specifically glucose are good for us, providing energy for our brains and red blood cells, but when over-consumed, that same sugar then becomes detrimental to our bodies in various ways. And this concept applies to all nutrients to one degree or another — when consumed in their proper amounts and ratios, nutrients are beneficial, but when not, they trend towards detrimental.

So you should balance your nutrients, most of all the macronutrients, the proteins, carbs and fats, so that you consume them in specific ratios to each other. Within protein, you should balance the ratio of "complete" to "collagen" protein. Within fats, you should balance the ratio of various fats to each other, including saturated, monounsaturated and polyunsaturated, as well as the ratio of omega-6 to -3. And within carbs, you should balance the ratios of starch to fruit while also in some cases combining various types of carbs in one meal to balance your blood sugar. Additionally, you should balance the ratio of macrominerals in your meals, especially the relationship of potassium to sodium/chloride (salt), as well as magnesium to calcium and phosphorous.

Additionally, this diet includes lots of fiber, with a balance between soluble and insoluble or fermentable fiber, as well as a diversity of phytonutrients abundant in most types of plants.

Premise 7: Taste and Satiation

Your diet should be delicious and most of all satiating, especially given that multiple studies and anecdotal evidence suggest that if people do not like their diet, they do not follow that diet. We modern Americans are perhaps conditioned to think that nutritious foods should taste bad, like steamed broccoli and tofu, while non-nutritious foods should taste good. But this conditioning is flawed because evolution designs us so that our pleasurable tastes direct us to nutritious foods and unpleasurable tastes direct us away from non-nutritious or toxic foods. By this reasoning, you should use your sense of pleasurable taste to steer you towards the most nutritious foods. But in our modern world, we encounter two problems with this reality. For starters, we have lost some of the art of making nutritious foods delicious through processing, cooking and combining ingredients. At the same time, food companies have tricked our taste buds into tasting non-nutritious foods as delicious.

Along those same lines, we are also possibly conditioned to believe that nutritious foods are not satiating. But in reality, when eating nutritious foods in the right ways, our body sends signals to our brain that we are satiated because our body, indeed, received the nutrients necessary for its functioning. Most Americans overeat because they do not receive the nutrients they need in their diet, and thus keep overeating.

So, in the end, this diet recommends that you make your meals delicious and satiating using mostly ancient and even current culinary practices for combining and preparing whole foods. At the same time, even when you merely follow the premises already mentioned above, you are more likely to make your meals more delicious and satiating—that is, you should avoid fake foods and supplements; prepare your foods to optimize their digestion while eating whole plants; prepare your foods also to reduce antinutrients, toxins and storage proteins; combine your foods to contain all nutrients in the proper ratios. If you then show even some slight culinary talent in combining flavors, you will, indeed, eat well and love your food, hopefully in the company of people you like and love.

EXAMPLE OF METHODOLOGY

As noted before, I use the four methodologies to analyze these seven premises. When doing this, I look for the similarities between the four methodologies to determine if one is supporting the other. However, I also look for the discrepancies and then attempt to explain those discrepancies when possible. After drawing my conclusions, I then make my recommendations for the optimum diet and as mentioned before, show how you can modify those suggestions for your own needs, such as weight-loss, anti-aging, disease prevention, as well as mental and physical performance.

To explain this approach better, let's consider an example that is now popular in the media: whether collagen provides benefits to our health. Through nutritional physiology, I reveal that collagen is safe even in high amounts, almost completely digestible in many forms, as well as thoroughly utilized by various tissues. I also reveal that humans are made of about 30% collagen protein and 70% complete protein.

In our modern diet, we eat plenty of complete protein as milk, eggs and meat but usually not enough collagen protein as broth, skin and certain types of sausages. Additionally, these two proteins, and most importantly, their ratios of amino acids, are radically different from each other. So while we can convert complete proteins into collagen proteins inside our body, as long as we consume enough Vitamin C, this conversion requires considerable work. However, over the course of our life, we cannot convert enough collagen in our body, so that if we are not getting enough in our diet, we start to deplete collagen throughout our body and trend ourselves towards, if not disease, then decrepitude.

Through ancestral consumption, I reveal that our ancestors, for tens of millions of years all the way until recently, always ate balances of complete and collagenous proteins, mostly because they consumed entire animals, including their collagen as skin, fascia, ligaments, tendons, nose, feet, hoofs and snouts. In many cases they melted those tissues into broth that became the foundation of soups and stews. But most moderns now eschew collagen altogether, focusing only on muscle meat, eggs and dairy.

Through nutritional establishment, however, I reveal that the establishment does not recommend collagen and in fact scores the protein at the bottom of the list on their various forms of protein rankings, like PDCAAS (Protein Digestibility Corrected Amino Acid Score).

Through scientific studies, however, I reveal that many scientists are now showing the benefits of collagen to our health—not just our skin, but to other, more vital parts of our bodies as well. Additionally, when collagen is combined with muscle tissues, like beef steaks, the quality of that protein improves overall. I also show some preliminary studies that suggest we should consume about ten to thirty percent of our protein from collagen and the remainder from completes.

In conclusion, three of these methodologies support the use of collagen, especially when combined with complete proteins. However, one methodology, nutritional establishment, does not. For that reason, I then investigate the establishment's methods for making their conclusions about collagen and conclude that those methods are flawed and left unexamined over more than fifty years. After some further analysis, I then recommend that we consume collagen in our diet, at under 30% of total protein, to support our joints, skin, hair, as well as other, more vital tissues that contain in collagen, including our cardio- vascular and nervous system.

As noted most modern humans consume little to no collagen in their diet because of these establishment recommendations and because these tissues, along with organ meats, also fell from favor over the past one hundred years. But at the same time, we also suffer inordinately from the diseases or maladies that afflict tissues our own collagen—all of which may explain bad posture, bad teeth, weak bones, wrinkled skin, painful joints, rigid bodies and possibly even heart disease and brain disorders.

PART 1

NUTRITIONAL PHYSIOLOGY

CHAPTER 1

OVERVIEW AND BASICS

In this section, I will discuss in greater detail the first of my methodologies, "nutritional physiology"—that is, the origin and form of nutrients and their function inside our bodies. Once you understand nutritional physiology, you can then apply the concept to the premises of your diet. Using simple and concise language, I explain how nutrients originate in our plant and animal foods, how they change when they are processed, cooked, chewed and digested. Finally, I show how we uptake and transport those nutrients and then utilize them for our metabolisms—for energy, growth and maintenance. We also recycle nutrients, convert them into other nutrients and store them for later use.

In the next section, I narrate the story of our ancestral consumption—that is, the evolution of our ancestors in relationship to food and nutrition. After that I then briefly explain the other two methodologies, establishment recommendations and scientific studies. In the final section I apply all four of these methodologies to the seven premises—and then draw my conclusions about the optimum diet for our various lifestyles. In the end, these premises are then summarized into several pages of information you can use to guide your diet.

Some people may struggle to understand the section ahead, especially if they lack any background in biology, nutrition or chemistry. For that reason, I recommend that you carefully read this section because to understand this book, and to understand your own body, you must understand your nutritional physiology: without that, you will find yourself susceptible to any trend diet or bad idea that comes your way. Additionally, I really streamlined the terminology and chemical pathways—and then summarized the most vital points at the bottom of each section. In the end, nutritional physiology, in this context, is not really all that hard to understand: it is the story of nutrients and their function inside our bodies in chronological order—in that sense, this is just the story of what happens when you eat food.

All animals and plants—and thus all nutrients—consist, almost entirely, of the four organic atoms—carbon, oxygen, hydrogen and nitrogen. They are the quintessence of life.

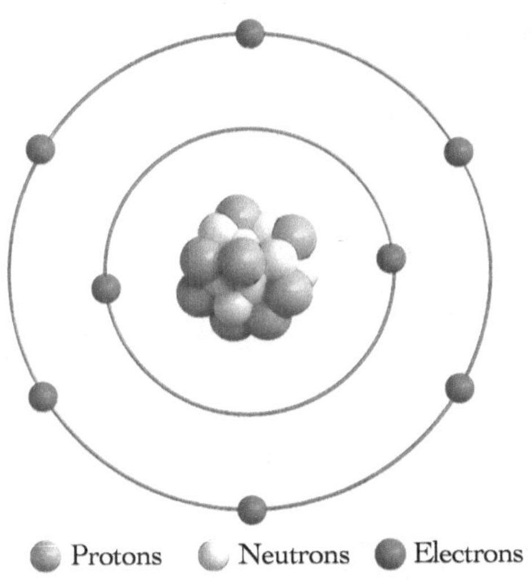

IMAGE 1.1- OXYGEN

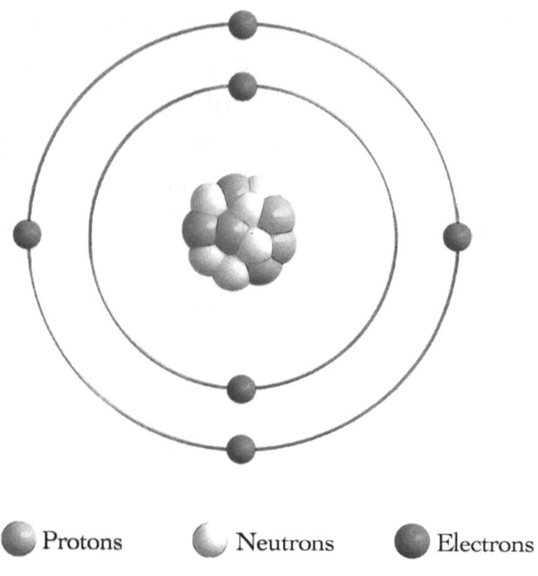

IMAGE 1.2- CARBON

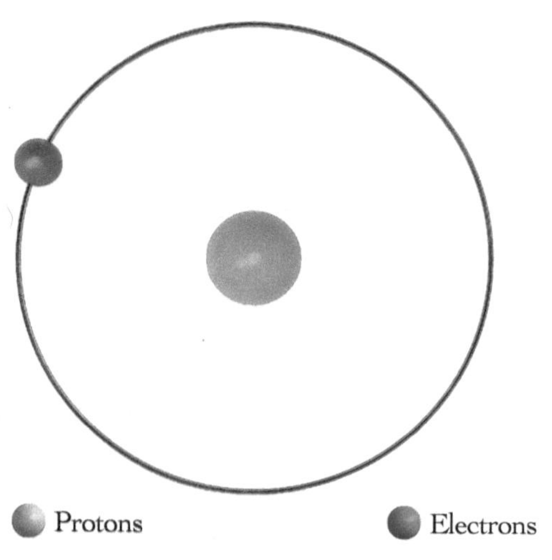

IMAGE 1.3- HYDROGEN

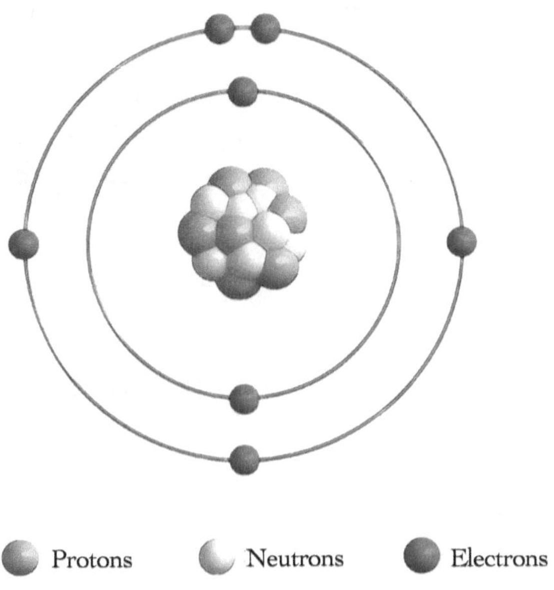

IMAGE 1.4- NITROGEN

These atoms combine to form the three macronutrients of life, called sugars, fatty acids, and amino acids. Sugars, such as glucose, are just made of carbon, oxygen and nitrogen.

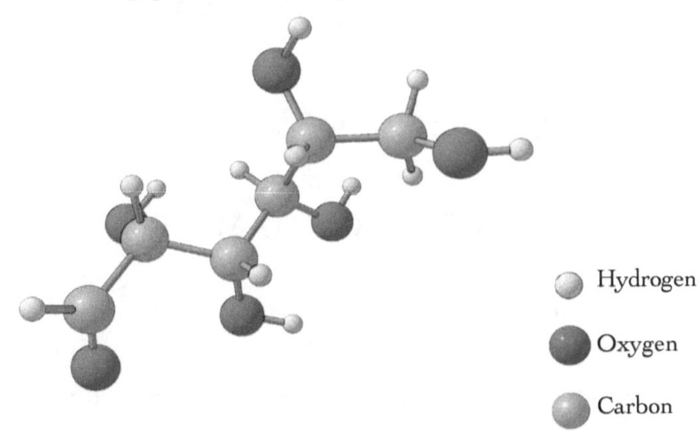

IMAGE 1.5- GLUCOSE

Fatty acids, including one of the main ones, palmitate, are also made of these same three atoms.

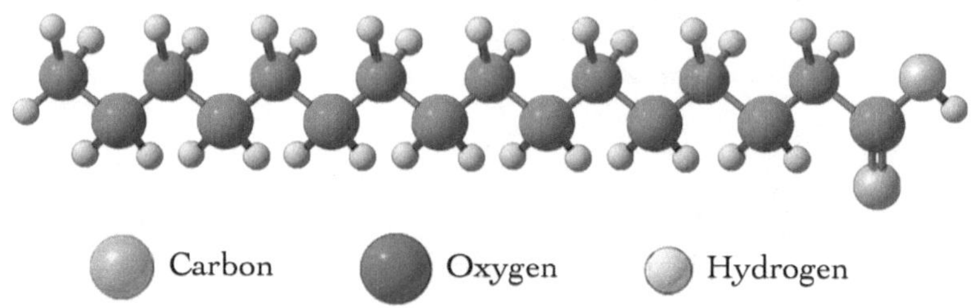

IMAGE 1.6- PALMITIC ACID:
SATURATED FATTY ACID

Amino acids are made of these three atoms, as well as nitrogen. Some amino acids also contain another atom, sulfur.

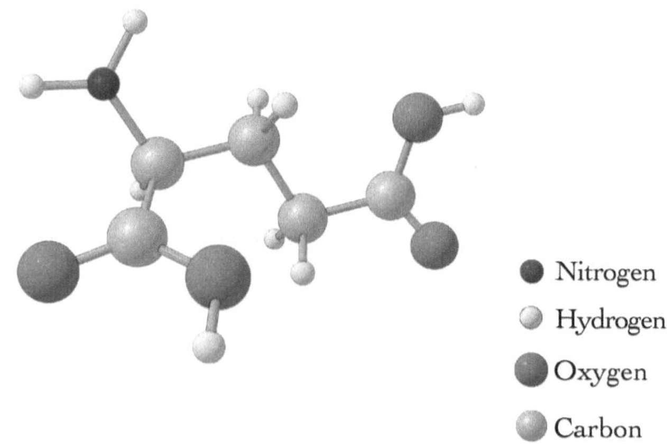

IMAGE 1.7- GLUTAMATE: AMINO ACID.

Humans convert most of these macronutrients from one form into another inside their bodies, mostly in the liver, which is made easier because they all contain the same or nearly the same atoms and are sometimes in similar constructions. These macronutrients then form together to form larger molecules: amino acids into proteins; sugars into carbohydrates; and fatty acids into fats. All organisms, single-cells, fungi, plants and animals, consist almost entirely of these macronutrients.

We also consist of other atoms, typically the minerals that are further categorized as macrominerals and trace minerals. The seven macrominerals are potassium, magnesium, phosphorous, calcium, sodium, chloride and sulfur, with six of these defined as electrolytes. We also consist of trace minerals mostly as parts of enzymes, including chromium, cobalt, copper, fluorine, iodine, iron, manganese, molybdenum, selenium, and zinc. Interestingly, although our universe contains over one hundred known atoms, we are only made of about twenty of them.

If we are melted down to our atoms, so to speak, humans are about 65% oxygen, 15% carbon, 10% hydrogen, and 3% nitrogen. Due to our bones, we consist of about 3% calcium and phosphorous, and the remaining three or 4% is the minerals listed above. On a larger scale, humans are also about 61% water; 16% protein or their smaller molecules called amino acids; 16% fat or their smaller molecules called fatty acids; and only 1% carbohydrates or its smaller molecules called sugars. The final 6% is minerals.

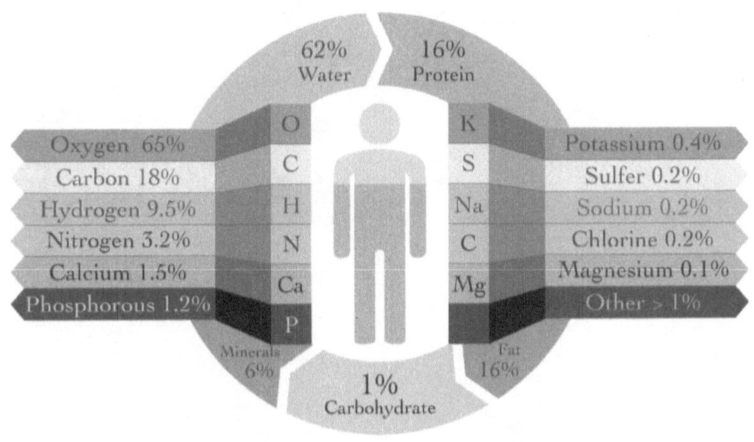

IMAGE 1.8- HUMAN, DIVIDED INTO ATOMS AND MACRONUTRIENTS

In this section I will mostly focus on the macronutrients—protein, carbs, and fats—because they play the largest role in our diet. "Nutritional physiology" is about these macronutrients and the two types of metabolism they generate inside our bodies. One metabolism, called catabolism, uses glucose and fatty acids to generate energy for our bodies in the form of ATP. The other metabolism, called anabolism, uses mostly amino acids to make many of the parts of our bodies.

CHAPTER 2

PROTEIN

FORM

All proteins consist of at least twenty or twenty-one types of amino acids. But proteins differ from each other in several ways. First, they contain different amounts of these types of amino acids, ranging from 100 to 1000 or more. Second, proteins contain different ratios of all those types of amino acids. Third, proteins connect all those amino acids in different ways and finally fold them into different shapes that are usually three-dimensional. So with all these variations, we humans now contain about one hundred thousand different proteins in our bodies, all made from just twenty-one amino acids.

In the human food chain, most amino acids originate in plants. From sunlight, water and air, plants photosynthesize glucose and from that one molecule make twenty of the twenty-one amino acids that the plant then uses for various functions. Animals, of course, then eat the plants, as well as other animals, to attain those amino acids. Additionally mammals, including humans, can make some amino acids inside their bodies from precursors; these are deemed by nutritionists as the "non-essential amino acids," but we cannot make others, called the "essential amino acids." However, I do not believe those distinctions are all that important to our diet, so I only mention them occasionally.

Plants and animals use amino acids in some of the same ways, but also in very different ways, and accordingly, they produce amino acids in different ratios to each other. For this reason, animal proteins are generally deemed "complete" by nutritionists because they contain amino acids in ratios that work best in our own animal bodies, while plant proteins are deemed "incomplete" unless at least two plants are paired together to make them more complete.

Humans across the globe generally acquire their proteins from animal foods, such as eggs, dairy, meat and various forms of collagen, perhaps broth most of all. Many acquire considerable amounts of their protein from plant floods, including from combining grains and legumes or nuts.

DIGESTION

We digest proteins in two ways: through pre-digestion, which happens before we put food in our mouths, and digestion, which happens once the food is put into our mouths. During predigestion, we cut or grind our proteins into smaller pieces, which breaks some of their bonds and exposes more of their surface to digestive chemicals later on. We can also predigest plant proteins through fermenting or sprouting. We then denature our proteins through cooking and other techniques, such as curing, smoking or soaking in acids: these processes unfold proteins and flatten them, making the bonds between amino acids easier to cleave later on during digestion. So, before we even put the protein into our mouth, we have already predigested that protein—which helps to make our digestion more complete and efficient later on.

During digestion, once we put the food into our mouth, we chew the protein into even smaller pieces that then pass into our stomach, where our hydrochloric acid activates the enzyme, pepsin, which begins to cleave the bonds between the amino acids: the acid also kills any pathogens. From there the protein flows into the small intestine, where the pancreas also releases several other enzymes that further cleave the amino acids. Once the protein is dissolved into either just amino acids or peptides, those compounds are then absorbed by the villi—tiny fingers that line the wall of the intestine—which then transport these compounds into our blood.

Because of the presence of fiber and antinutrients, plant proteins do not digest as completely or efficiently as animal proteins. However, through proper pre-digestion, such as soaking, sprouting or fermenting, and cooking, you can likely make plant proteins digest nearly as well as animal proteins.[7]

FUNCTION AND ANABOLISM

Once in our blood, most amino acids flow freely, while some are bound again into proteins. But they all flow to our liver, where they are then redistributed back into our bloodstream. Once arriving at our cells, the amino acids are then taken up into the cell and mostly transported to our ribosomes, where they are anabolized—that is, the amino acids are reconstructed back into larger molecules like proteins. More specifically, our DNA codes templates called RNA, which are then transported to the ribosomes, and then the RNA constructs those amino acids back into proteins, which are then delivered around the cell for various purposes.

Generally, our cells anabolize amino acids into various forms with different purposes. In rare cases amino acids are solo in the body, such as the neurotransmitter, glutamate. But in almost all cases they are bundled into peptides, enzymes and most of all proteins. Peptides are the smallest of these molecules, always under fifty amino acids altogether, serving mostly functional purposes as hormones and neurotransmitters. Enzymes are the second largest of these molecules, containing somewhere between fifty and one hundred amino acids while usually combined with minerals, such as magnesium and zinc, to help in the deconstructing and constructing of other molecules in our body. But nearly all amino acids in our body are bound into the larger molecules of protein.

Despite the enormous amounts of types of protein in our body as mentioned above, mammals, including humans, contain mostly two broad classes of proteins that coincide with the largest tissues in our body: complete proteins, the ones already mentioned, are abundant in our muscles and organs; and collagen proteins abundant in our connective tissues such as bone, ligaments, tendons, fascia, skin, etc. All of these complete proteins are similar to each other in their ratios of amino acids. And collagen proteins are also quite similar to each other and to another class of proteins called elastin that makes our muscles more flexible. In our makeup, we humans are about 70% complete proteins and 30% collagenous proteins.[8]

	Muscle	Liver	Collagen	Elastin	Keratin
Alanine	6.6	6.5	8.1	11.1	4.2
Arginine	6.1	6.3	8.4	10.1	5.2
Asparagine	4	4.2	0	0	0
Aspartic Acid	10.1	10	6.6	6.5	7.8
Cysteine	1.4	1.3	0.6	1	7-3
Glutamine	4	4-3	0	0	0
Glutamic Acid	15.2	15.8	12.4	13.2	14
Glycine	7-5	7.2	20.6	18.5	3.8
Histidine	3-5	3.2	0.8	1.2	1.8
Hydroxyproline	0	0	11.4	10	0
Isoleucine	5.2	5	1.5	2.8	3
Leucine	8	8.5	2.9	5-5	6.8
Lysine	9	9.2	3-4	4	6.3
Methionine	2.5	2.3	0.6	1.1	1.9

	Muscle	Liver	Collagen	Elastin	Keratin
Phenylalanine	4-5	4.7	2.1	3-3	4-4
Proline	5	4.8	11.5	12	9-4
Serine	4.2	4	3-4	4.5	3.9
Threonine	4	4-4	1.9	2.2	2.6
Tryptophan	1.2	1.1	0	0.5	1.3
Tyrosine	3.8	3.6	0.5	1.8	2.4
Valine	6	6.2	2.4	3-9	4-5

TABLE 2.1- COMPARISON OF AMINO ACID RATIOS IN MUSCLE, LIVER, COLLAGEN, ELASTIN, AND KERATIN PROTEINS.

Note that muscle and liver have similar ratios; and collagen, elastin and keratin have similar ratios. But these two groups have very different ratios from each other.

Once constructed many of these compounds are short or long-lived. Enzymes and peptides are frequently destroyed within milliseconds of their creation, while proteins endure for anywhere from seconds to years. But when these compounds are destroyed, some of their amino acids are left intact and thus recycled back into the body. However, about 20% of the amino acids are damaged, through oxidation, glycation, viruses or bacteria and thus are discarded. Humans, amazingly, turn over about 300 to 400 g of protein per day on average[9]—the equivalent of about 60 eggs—that is, we destroy that amount of protein, and then replace that amount. So we need to generate about 300 g of protein per day, but about 80% those amino acids come from recycled amino acids and only about 20% from our diet.[10] In this way we create, maintain and destroy ourselves every day just like the Hindu triumvirate of Brahma, Vishnu and Shiva are doing with the universe in Hindu theology.

AMOUNTS AND RATIOS

So we eat protein every day to replace discarded proteins. If we do not consume enough protein for this purpose, we start to encounter some problems—that is, we slow our anabolism or break down our muscles and pull amino acids from them. If we consume too much protein, we then cannot utilize those amino acids and thus we convert them into glucose and ketones. So generally we need to consume only enough protein to replace damaged protein—not more or less.

However, we not only need certain amounts of amino acids; we also need them in certain ratios to each other—that is, in ratios that match complete proteins as well as collagenous proteins. When we consume proteins in these ratios, we better anabolize those tissues. But when we consume our amino acids in improper ratios, we in some cases cannot utilize some of them, so we must convert them. Or in other cases, we do not acquire the amino acids we need so we pull them from our muscles or make them from precursors Presumably, this slows down our anabolism, especially over our lifetime, and probably trends us towards disease or at least decrepitude.

Generally, throughout this book, I argue that humans in most cases should consume not only protein but all nutrients in the amounts and ratios that we use them—not more or less—because that then makes our metabolism, both anabolism and catabolism, more efficient, which results in many benefits over our lifetime. When we consume more protein than we need, we thus digest, transport, absorb and then convert nutrients that we do not need—and this is inefficient as all these processes require time, tissue and metabolism in our body. We thus expend all these resources without receiving any benefits. Through these conversions, we also release waste products, including uric acid, urea, and ammonia that are potentially toxic in higher amounts.[11,12]

Interestingly, all humans need about the same amount of protein per day per lean mass. If you carry more muscle and organ tissues, you of course need more protein; and the reverse. This is, more or less, determined genetically not only for humans but most animals. However, when we exercise more, we need more protein per day because we are damaging more amino acids; and we tend to build more muscle mass.

Throughout this book, I return to this topic repeatedly: when consuming nutrients in their proper ratios, we optimize our metabolisms; when in their improper ratios, we compromise our metabolisms—which trends us towards detriments, including disease. As explained in a later section, we evolved in this direction. With that being said, however, we always convert some amount of nutrients inside our bodies all the time.

SUMMARY

From this short tutorial on the nutritional physiology of protein, you have already learned the basics of protein that you likely never heard of, which will help you understand your optimum diet.

- You should prepare your protein to optimize digestion—that is, to make your digestion more complete and more efficient (premise two).

- If you consume animal protein, you do not need to worry about antinutrients and toxins. If you consume plant proteins, you should prepare them to reduce these compounds (premise three).

- You should also combine your plant proteins to make them more complete (premises four and six).

- You should eat enough protein to replace damaged and discarded amino acids.

- If you eat too little protein, you suffer detriments, including reduced anabolism, muscle loss and other detriments I discuss in later sections (premises four and six).

- If you eat too much protein, you suffer detriments. You waste metabolism on digesting, transporting, absorbing and converting nutrients you do not need while generating more waste products that are potentially toxic (premises four and six).

- If you eat your protein with improper ratios of amino acids, you suffer detriments, especially over your lifetime. This generally happens from eating too many grains, like wheat, without legumes or nuts; or from eating too many complete proteins, especially muscle meat, without enough collagen proteins (premises four and six).

- If you exercise more, you need more protein, but not necessarily much more, because exercise damages more amino acids, which you then must replace. You also need more protein to build stronger muscles.

- And why not make your proteins tasty with some culinary finesse in the prepping, cooking and flavoring of your foods (premise seven).

CHAPTER 3
CARBOHYDRATES

FORM

There are three types of sugars: glucose, fructose and galactose, which are chemically similar. Fructose tastes the sweetest to humans, followed by glucose, then galactose. We humans do not metabolize two of these sugars, fructose and galactose, so we must convert them into glucose first — so in that sense, they are faux or imitation sugars. Additionally, these two sugars are more toxic than glucose, implicated in both glycation, oxidation, obesity, senescence, and fatty liver disease.[13] For that reason, our bodies generally convert them into glucose as soon as possible.

During photosynthesis plants create glucose and, in some cases, convert that into fructose and occasionally into galactose. They then catabolize glucose to generate their own energy or ATP. But they also bond glucose and the other sugars to make larger compounds called carbohydrates, which are used by the plant for several purposes. When bonding glucose to fructose, plants form the carbohydrate, sucrose, which they place in their fruits, berries and flowers; this then attracts animals to eat the sugar in these foods but not digest the seeds, which are then excreted down the way in piles of manure — so in that sense, plants form fruits for reproduction. Sucrose is also the carbohydrate in honey, syrups and table sugar.

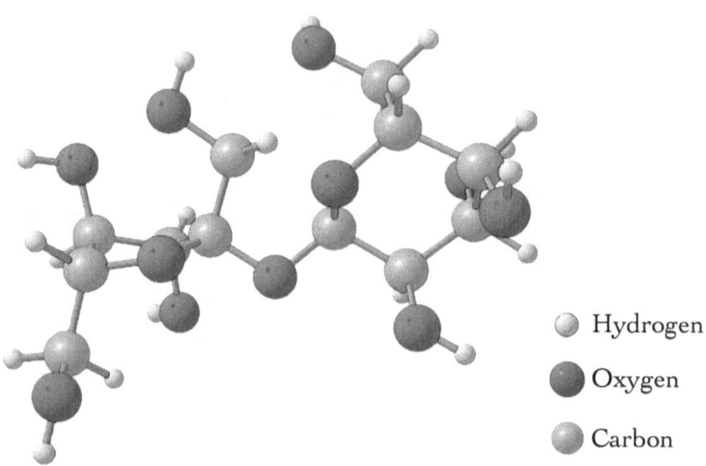

IMAGE 3.1- SUCROSE

When plants bond glucose to glucose, they form starch, which is then stored in tubers, like potatoes, which the plants use later during times of cold and drought. They also store starch in their seeds — which are then used by the seeds themselves for energy before they break through the ground and start to photosynthesize their own sugar.

Some starches are small, such as maltose (the carb used to brew beer), which consists of only two molecules of glucose, while other forms of starch contain hundreds of glucose molecules. Plants also make fiber from glucose and other sugars; they then use this fiber to form the structures of their body, such as their roots, stems, limbs, etc., and to form barks and husks that protect them from their environment.

When not receiving enough glucose in the diet, humans also make glucose inside their bodies from certain amino acids (the glucogenic amino acids) as well as another molecule called glycerol (the molecule that bonds fatty acids into triglycerides, which I will cover in the next section).[14] We humans also store glucose in our liver and muscles as glycogen. Human females also bond glucose to galactose to form lactose, which is the carb in human and dairy milk.

PREDIGESTION AND DIGESTION

We humans specialize in consuming the stored carbs found in fruits, tubers, and grains, primarily to provide fuel for our larger brains. As noted before, plants design their fruits and berries so animals will eat them and, accordingly, we humans have consumed both of them raw for millions of years—even though we can cook them as well. However, plants want to protect their roots and seeds, so these foods, especially grains, tend to come with antinutrients, toxins and storage proteins. Accordingly, our ancestors predigested these foods through soaking, fermenting, sprouting, grinding or mashing—which deactivates many of these compounds to one extent or another. We then cook them, causing the carbs to break down into smaller chains like maltose and swell with water and gelatinize—which makes them easier for us to digest. In more recent times, we also remove nearly all the fiber from our carbohydrates, making such foods as white flours, fruit juice and sweeteners like table sugar.

When eating carbs, we first slather them in our mouths with saliva and amylase, the enzyme that breaks down starch into smaller chains of sugar. We do not digest carbohydrates in our stomachs, but once they pass into our small intestine, they are slathered with more amylose and other enzymes from our pancreas that target various carbs. All carbs are rendered into smaller and smaller units, until just the individual sugars remain, including glucose, fructose and galactose.

FUNCTION AND CATABOLISM

Once absorbed into our blood, these three sugars flow to our liver. Fructose and galactose are converted into glucose and then usually stored for later use; or fructose in particular is converted into the fatty acid, palmitate, which can trend us towards fatty liver disease. However, most glucose merely passes through our liver back into our bloodstream, where our cells absorb the glucose to make ATP through two cycles of catabolism: glycolysis and the Krebs Cycle, which continues through the Electron Transport Chain. Glycolsis happens in the pool of liquid inside the cell called the cytoplasm. While passing glucose through multiple chemical steps, glycolysis uses ATP but in the end produces more ATP, so that one molecule of glucose renders two molecules of ATP. In the process, though, glycolysis also produces another molecule, called pyruvate, which then releases even more ATP downsteam in the Kreb's cycle.

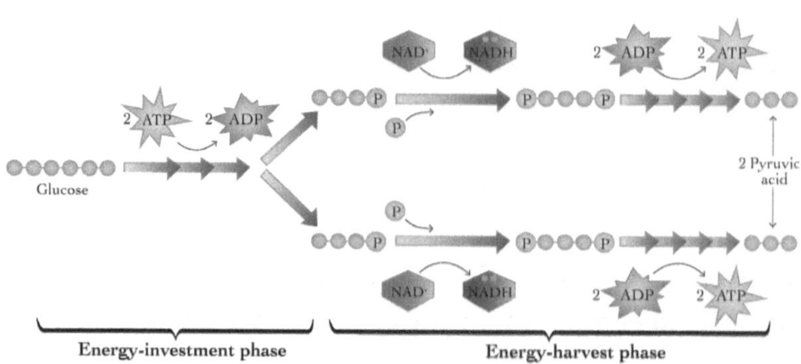

IMAGE 3.2- GLYCOLYSIS: GLUCOSE CONVERTED INTO ATP.

During extreme exercise such as running and sprinting, when our cells are low on oxygen, the pyruvate enters "anaerobic glycolysis" to generate small amounts of ATP, as well as another molecule called lactate (usually called lactic acid), which is also used to generate ATP. But our cells can only endure this state for short periods of time before they start to die. However, in almost all situations, our cells are saturated with oxygen and, in that case, the pyruvate then flows to the powerhouse of the cell, the mitochondria, where it enters into the Krebs Cycle to produce six more ATP from two pyruvates and then into the Electron Transport Chain where twenty-six or so more ATP are generated. Altogether, one molecule of glucose generates about thirty-four ATP.

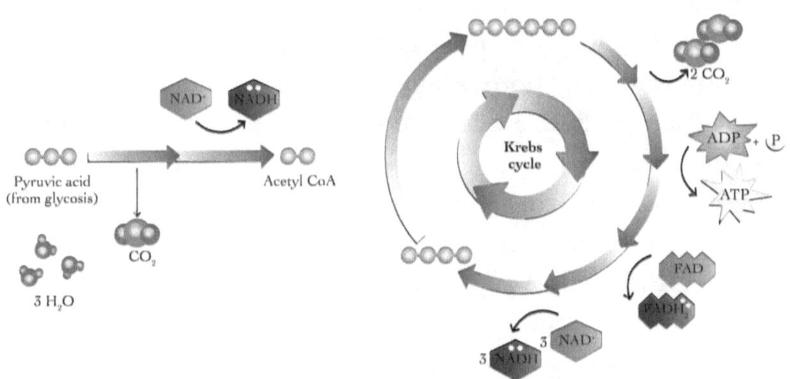

IMAGE 3.3- THE KREBS CYCLE.

As explained later, all fatty acids are also converted into acetyl-CoA.

AMOUNTS AND RATIOS

Only some tissues prefer carbs for catabolism, including our nervous system (brain, eyes, nerves and etc), our red blood cells and some minor tissues as well; but those tissues only need about 150 to 200 g of carbs per day on average-the equivalent of around six slices of whole wheat toast. Our brains catabolize glucose at the same rate on average per day so, whether you are sofa surfing or writing the great American novel, you are still catabolizing the same amount of glucose, and you cannot increase your brain power by eating more carbs. Our muscles and organs, while preferring fats for catabolism, also uptake excess glucose from the blood—but then revert back to their preferred fuels as soon as possible. However, at higher levels of exercise, such as with running or tennis, our muscles prefer glucose—but for that we only need about fifty additional g of carbs per hour at the most.

Our brains must have glucose for fuel. So, when under-consuming carbs, we merely make glucose inside our bodies from precursors, such as amino acids and glycerol—the process known as glucogenesis. At the same time, our body makes ketones from amino acids—the process known as ketogenesis; and then our brain uses these compounds for energy as well. Additionally, our brains catabolize acetic acid, which comes from fermented food, including alcohol, as well from our colon. So in short, when not consuming enough carbs, we merely make glucose and ketones inside our bodies—which does not provide any known benefits.

When over-consuming carbs, especially ones with their fiber removed, we absorb them too quickly, which first spikes and then lowers our blood sugar. We become hungry again too quickly and then over-consume carbs again. During these spikes in blood sugar, we are more likely to generate damage from both oxidation and glycation. Also, our organs and and muscles absorb the glucose to lower our blood sugar—even though these tissue prefer fats.

If our blood sugar remains high after all of this, we then convert glucose into glycogen and store that in our liver and muscles for later use, but only enough to last about one day. But our glycogen stores are usually close to full, so we actually convert most of the excess glucose into the fatty acid, palmitate, making ourselves fatter. Generally, we trend ourselves towards the diseases of civilization when overeating carbs, especially refined ones.

FIBER

As explained before, plants make fiber from glucose and, in some cases, from fructose and galactose — so it is defined as a carbohydrate. All plants contain fiber but certain plants in our diet are more abundant in fiber, including fruits, tubers, and seeds. We cannot actually digest fiber into its underlying glucose, so it passes through our stomach and small intestine intact, then eventually flows into the colon.

Generally, we categorize two types of fiber: insoluble and soluble. We receive about equal ratios of these fibers when eating diverse diets consisting of whole plants. Insoluble fiber does not mix well with water, but nonetheless binds with our food and slows down the digestion and passage of our food; it also binds with cholesterol, preventing its uptake into our blood. Once arriving in the colon, these fibers are then passed with our stool.

Soluble fiber also binds with our food, slows down digestion, draws in more liquid and, paradoxically, speeds the passage of food through our gut. Once the fiber arrives in our colon, our microorganisms emit enzymes to dissolve the fiber into glucose; they then catabolize the glucose themselves for energy. And in the process, these microorganisms then liberate nutrients from the fiber that flows into our blood for our own use, including minerals, phytonutrients, antioxidants and other compounds.[15] At the same time, our microflora generates many benefcial compounds, including B-vitamins, neurotransmitters, molecular hydrogen (one of our strongest and most effective antioxidants), other gasses and many other compounds. But most importantly, our microflora generates the three short-chain fatty acids, already mentioned, that provide fuel for various forms of catabolism throughout our body. One of those short chains, butyrate (butyric acid), provides specific fuel for the cells lining our colon — thus keeping our colon healthy. Another one, propionate (propionic acid), flows through the blood to the liver, where it is converted into glucose. And the final, acetate (acetic acid), (also found abundantly in vinegar and other fermented foods) is catabolized by most cells in our body, while providing many benefits along the way, including greater catabolism of fat, resulting in weight loss, as well as decreased blood sugar, increased insulin sensitivity, and improved cholesterol.[16] Some studies suggest, too, that our brains prefer acetic acid for catabolism, perhaps even over glucose, under certain situations.[17] Similarly, studies show that soluble fiber in general improves the health of our brain, resulting in better concentration, performance and mood while reducing depression and anxiety.[18,19]

Plants generate many types of fiber, and our colons contain hundreds of different types of micro-organisms that selectively feed only on some of those fibers and then produce only certain of the nutrients listed above. For example, some microorganisms feed on certain fibers and generate butyric acid, others acetic acid, and others molecular hydrogen. So, when consuming more varieties of plants in our diet, with more variation in fiber, we then produce greater ranges of nutrients which then enhances our health.[20]

If we under-consume fiber, we lose all the benefits above; when eating our carbs without fiber, we risk many potential detriments already mentioned. If we over-consume fiber, we risk slowing our digestion too much, making our meals unpalatable and overfilling our stomachs too much, thereby preventing us from eating enough macronutrients.

SUMMARY

So, from this short tutorial, you have already learned the basics of carbs to help you design your own optimum diet.

- You can eat your fruits and berries raw or cooked (premise 2 and 3).

- You should eat your carbs whole—that is, whole grains,

fruits and tubers — because the fiber and other compounds provide many benefits (premise four).

- You should thoroughly predigest and cook your starches, especially your grains (premises 2 and 3).

- You only need about 150 g per day of carbs on average for your brain and your blood cells (premise six).

- You cannot increase the performance of these tissues by consuming more carbs (premise six).

- If you exercise more intensely, you need more carbs, but only about fifty additional g per one hour of this type of exercise (premise 6).

- If you undereat carbs, you then just make them inside your body from precursors, which does not provide any advantages and only disadvantages, which I will discuss in the pages ahead (premise six).

- If you overeat carbs, especially with their fiber removed, you tend to digest your carbs too fast, spike and then sharply lower your blood sugar, causing you to become hungrier sooner, and then overeat carbs again (premise six).

- At the same time, you tend to cause many detriments, including glycation, oxidation and lipogenesis, while making it harder for you to catabolize fats, which para-doxically can make you fatter (premise 6).

- If removing fiber from your carbs, you tend to overeat them, causing all the detriments above while also losing all the additional benefits and nutrients that fiber pro-vides (premise 4).

CHAPTER 4

FATS

FORM

Fats consist of smaller molecules called fatty acids that are shaped like snakes. They all have the same type of head as well as tails, known as "chains", made from carbon with hydrogen on both sides. However, fatty acids differ from each other in several ways as well. First of all, they differ in the length of their chains which are either defined as short, medium or long. Secondly, they differ in the shapes of their chains, which are either straight, bent or curved. Thirdly, fatty acids differ in their degree of saturation, which in turn makes them stiffer or more liquid at room temperature. Saturates are most stiff and thus solid at room temperature; their chains are always straight but vary in length from short to medium to long, and this in turn makes them more stable and resistant to oxidation or damage.

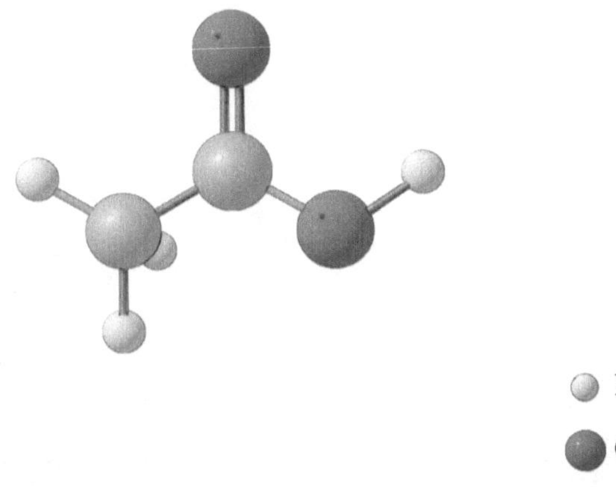

IMAGE 4.1- ACETIC ACID: SHORT-CHAIN SATURATED FATTY ACID
Low sensitivity to oxidation.

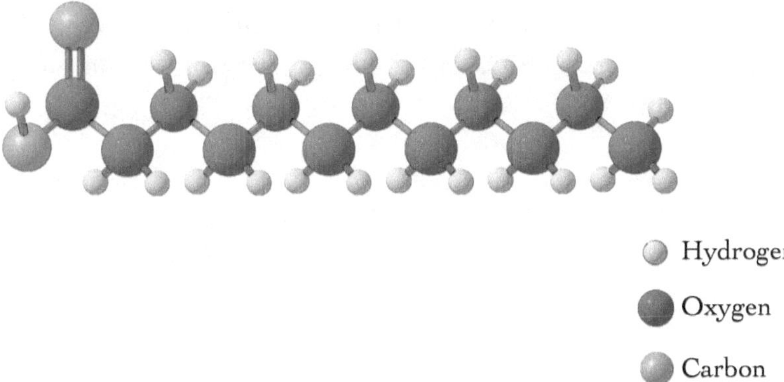

IMAGE 4.2- LAURIC ACID: MEDIUM-CHAIN SATURATED FATTY ACID.
Low sensitivity to oxidation.

Monounsaturates are liquid at room temperature. They only contain long and bent chains, and they are slightly sensitive to oxidation.

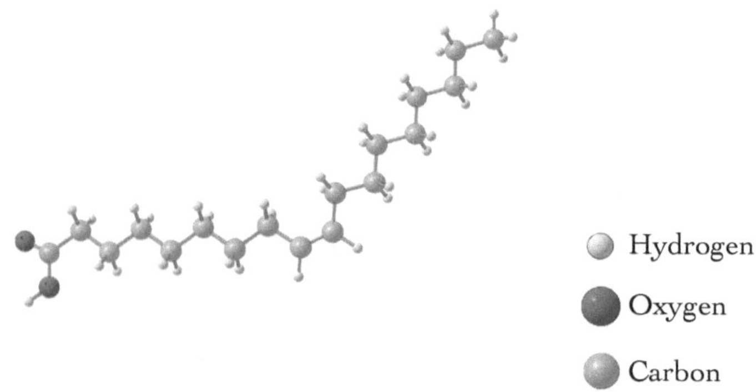

IMAGE 4.3- OLEIC ACID. LONG-CHAIN MONOUNSATU-
RATED FATTY ACID WITH A BENT TAIL.
Medium sensitivity to oxidation.

Polyunsaturates are liquid even in the refrigerator and highly sensitive to oxidation, including inside our bodies. Polyunsaturates, specifically, are further defined into two categories of omega-6 and -3. While all polyunsaturates are long chains, some have shorter chains than others, and thus I refer to the ones that predominate in plants as shorter chains, and the ones that predominate in animals as longer chains.

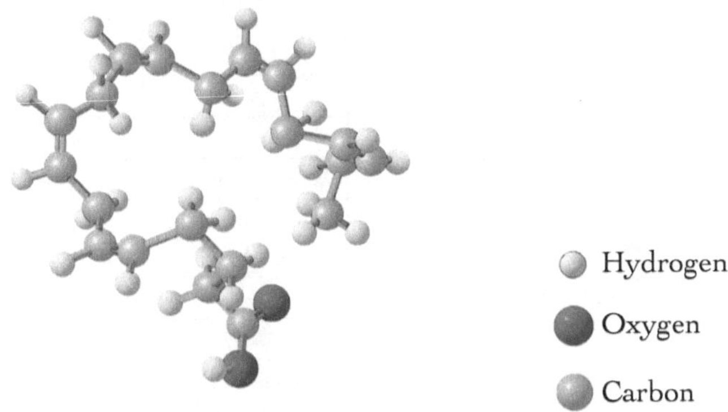

IMAGE 4.4- EICOSATETRAENOIC ACID: LONGER-CHAIN POLYUNSATURATED FATTY ACID, OMEGA-3.
High sensitivity to oxidation.

Nature contains about fifty fatty acids altogether, somewhat equally distributed amongst the three categories. However, we humans and most other animals only regularly consume about ten to fifteen of them in significant amounts and nearly all of these are the long chains from all three classes of saturation.

Type	Isomer	Name	Note
	18:1 undiffer-entiated	oleic acid	Abundant in olive oil, avocado, nuts, and animal fats and palm oil
	18:1 c		
	18:1 t		
	20:1	gadoleic acid	
	22:1	undifferentiated erucic acid	
	22:1 c		
	22:1 t		
	24:1 c	nervonic acid	
Type	Isomer	Name	Note
Polyunsaturated Fats	16:2 undiffer-entiated	hexadecadienoic acid	
	18:2 undiffer-entiated	linoleic acid	Shorter chain Omega-6. Generally higher in plant fats and lower in animal fats. Converted into the longer chain Omega-6 arachidonic acid
	18:2 n-6 c,c		
	18:2 c,t		
	18:2 t,c		
	18:2 t,t		
	18:2 i		
	18:2 t not further defined		
	18:3 undiffer-entiated	linolenic acid	
	18:3 n-3 c,c,c	alpha-linolenic acid	Shorter chain Omega-3. Generally low in plant and animal fats. Converted into the longer chain Omega-3 Eicosapentaenoic acid then into Docosahexaenoic acid
	18:3 n-6 c,c,c	gamma-linolenic acid	
	18:4 undiffer-entiated	parinaric acid	
	20:2 n-6 c,c	eicosadienoic acid	

Type	Length of Chain (tail)	Name	Note
Saturated Fats	2:0	acetic acid	Short chain : colonic fat
	3:0	propropioic acid	Short chain : colonic fat
	4:0	butyric acid	Short chain : colonic fat
	6:0	caproic acid	
	8:0	caprylic acid	Medium chain : found mostly in coconuts and dairy
	10:0	capric acid	Medium chain : found mostly in coconuts and dairy
	12:0	lauric acid	Medium chain : found mostly in coconuts and dairy
	13:0	tridecanoic acid	
	14:0	myristic acid	Long chain : found mostly in coconuts and dairy
	15:0	pentadecanoic acid	
	16:0	palmitic acid	Long chain : abundant in all animal fats and palm oil
	17:0	margaric acid	
	18:0	stearic acid	Long chain : less abundant in all animal fats and palm oil
	19:0	nonadecanoic acid	
	20:0	arachidic acid	
	22:0	behenic acid	
	24:0	lignoceric acid	

Type	Isomer	Name	Note
Monounsaturated Fats	14:1	myristoleic acid	
	15:1	pentadecenoic acid	
	16:1	undifferentiated palmitoleic acid	All monos are long chains
	16:1 c		
	16:1 t		
	17:1	heptadecenoic acid	

20:3 undifferentiated	eicosatrienoic acid	
20:3 n-3		
20:3 n-6		
20:4 undifferentiated	arachidonic acid	Small amounts in animal fats only, but abundant in the brain
20:4 n-3		
20:4 n-6		
20:5 n-3	eicosapentaenoic acid (EPA)	Abundant in marine fats and brains
22:2	docosadienoic acid	
22:5 n-3	docosapentaenoic acid (DPA)	
22:6 n-3	docosahexaenoic acid (DHA)	Abundant in marine fats and brains

TABLE 4.1- CHART OF FATTY ACIDS

Both plants and animals make fatty acids. Plants photo- synthesize glucose which they then convert into the long-chain saturate, palmitate, and then convert that into all other types of fatty acids. Animals, including humans, also convert glucose into palmitate, and then convert that into other fatty acids. Humans can make many fatty acids inside our bodies from other fatty acids, including nearly all saturates and monounsaturates. We can convert long-chain saturates into long-chain monounsaturates, and then vice versa. However, we cannot make the polyunsaturates from other fatty acids; these are called the essentials because we must receive them from our diet. However, we can convert shorter-chain polyunsaturates into longer-chain versions and then the reverse.

Once they make fatty acids, both plants and animals bond them to other fatty acids and other molecules to create larger compounds called phospholipids and triglycerides. Phospholipids are usually two layers of fatty acids stacked tail to tail, and both plants and animals use them to construct membranes around their cells and around the organelles inside their cells.

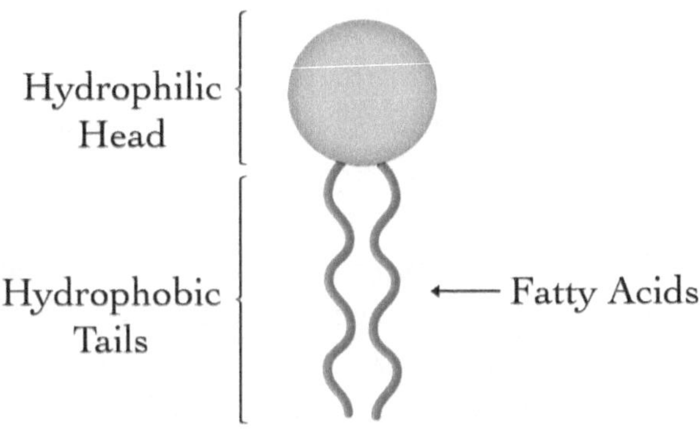

IMAGE 4.5- PHOSPHOLIPID

Triglycerides are three fatty acids bonded with another molecule called glycerol. Plants use triglycerides to store fatty acids, mostly in their seeds, such as nuts, to feed them while they are germinating. Animals, including humans, use triglycerides to transport fatty acids through their blood as well as store them as body fat.

We humans consume both the phospholipids and triglycerides from plants and animals. But we consume only small amounts of fats from phospholipids but enormous amounts of fats from triglycerides, including from plant oils and animal fats. These triglycerides all contain different fatty acids inside them; they are never just one fatty acid but combinations of various long-chains. Generally, the triglycerides in most animal fats contain about equal parts saturates and monounsaturates and smaller amounts of polyunsaturates from both the omega-6 and -3 families. Generally, most plant fats contain lower amounts of saturates, and higher amounts of monounsaturates and polyunsaturates: most of these plants contain more of the omega-6s, but others contain considerable amounts of omega-3. We mistakenly refer to animal fats as saturates because they contain enough saturates to become stiff at room temperature—even though they actually tend to contain more monounsaturates and polyunsaturates overall.

DIGESTION AND TRANSPORT

We predigest plant fats by grinding seeds into butters and extracting oils from plants like olives, palm fruits and various seeds. We also predigest animal fats by grinding them or rendering them for tallow and lard. But unlike proteins and carbs, fats do not predigest much during our usual forms of cooking; they mostly stay the same chemically. But there are exceptions to this rule: during prolonged heating, meaning for many hours, and possibly at higher temperatures, triglycerides sometimes dissolve into glycerol and free fatty acids.[21] At extreme heat for prolonged times, fats start to decompose and toxify through oxidation, isomerization or polymerization, which is why you should exercise caution with fried foods, especially ones that are commercialized.[22] So in short, we do not really predigest our fats much, if at all, during cooking, so we can eat them either raw or cooked; and indeed many cultures consume nearly all of their fats raw as plant oils, butter and seal and whale blubber without any detriments and possibly with benefits over cooked fats.

After swallowing fats and oils, we digest them by emulsifying or mixing them in our stomachs with water, and the same process then continues in the small intestines with the secretion of bile. All of this generates smaller blobs of fat with greater surface area for more exposure to digestive chemicals. Once fats arrive in our intestine, the pancreas then secretes the enzyme, lipase, to break them down into individual fatty acids and glycerol. After the fatty acids are absorbed into the cells of our small intestine, they are reconstructed back into triglycerides, absorbed into our blood, and then attached to much larger molecules, called lipoproteins, otherwise known as lipid rafts, that additionally contain proteins, cholesterol and fat-soluble vitamins, including vitamins A, D, E, and K. These lipoproteins are used to transport all these nutrients through the blood because fat and blood do not mix well with each other.

Once arriving at the cell, the lipoproteins, as well as the tri-glycerides inside of them, are deconstructed back into their smaller nutrients, which are then absorbed into the cell. But if our cells are sated with fatty acids and these other nutrients, the lipoproteins remain intact and are then transported to our liver, where they are deconstructed and reconstructed back into new lipoproteins and then sent back to our cells to undergo the same process all over again.

When consuming greater amounts of some saturated fatty acids, we generate lipoproteins with more cholesterol—which are called "bad cholesterol" because they supposedly contribute to heart disease. When consuming more mono and polyunsaturated fats, we generate lipoproteins with less cholesterol that then retrieve cholesterol from our cells, so we oddly call these lipoproteins "good cholesterol," which supposedly prevents heart disease. But in reality, these names do not make much sense because lipoproteins are not cholesterol; they merely contain different amounts of cholesterol.

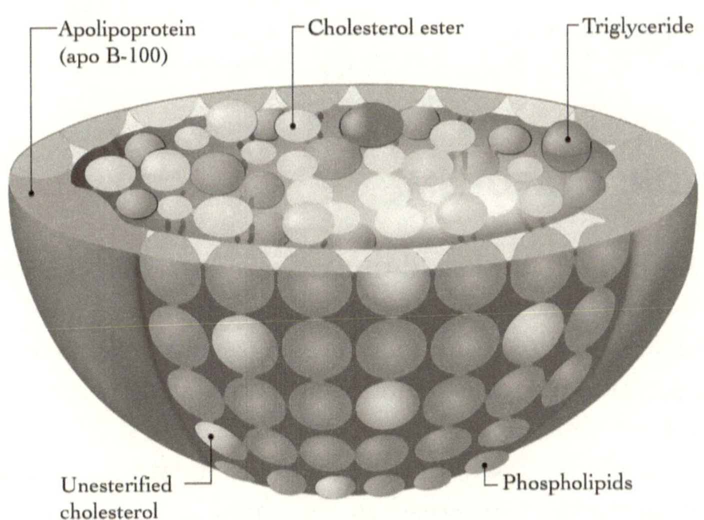

IMAGE 4.6- LOW-DENSITY LIPOPROTEIN:
"BAD CHOLESTEROL"

At some point, if our cells remain sated, the lipoproteins are then transported to our adipose cells, where they are deconstructed back into fatty acids, sometimes converted into other fatty acids, and stored for later use in body fat. Interestingly, all humans, as well as many primates, store fat with mostly the same ratios of fatty acids in them. In other words, if you consume nothing but dairy fat or olive oil, you still store your fatty acids in the same ratios with only slight variations due to diet, mostly in the omega-6s and -3s. These ratios are about 30% saturated fat, 50% monounsaturated fat, and 20% polyunsaturated fat. Generally, these percentages range up or down by about 5% for the class of fat.

So in digesting and transporting fat, we use an incredible amount of steps: with triglycerides in particular, we deconstruct them in our gut, reconstruct them in our blood, deconstruct them at our cells and reconstruct them again for storage—all in part because we cannot carry either single fatty acids or triglycerides through our blood because water and fats do not mix. Despite all these processes, however, and paradoxically, we use the least amount of energy to digest and utilize fats as addressed in other sections.

FUNCTION: ANABOLISM

Once absorbed into the cell, fatty acids are used for both anabolism and catabolism. Fatty acids are of course anabolized into phospholipids that are used to make membranes for our cells, as well as the organelles inside our cells. Our membranes, and their phospholipids, do not use short or medium-chain fatty acids at all, but only the longer chain versions with both straight and bent tails from all three classes of saturation.[23] Interestingly, we hold fatty acids in our phospholipids in approximately the same ratios as in our body fat.

Brain cell membranes require speed—so they use more polyunsaturates, especially as omega-3s, which, as noted, are more fluid but also more prone to oxidation.[24] Meanwhile our muscle cells need stability—so they use more saturates and monounsaturates which are more stable and less prone to oxidation. However, the differences are only minor. At the same time, phospholipids throughout our body are also sensitive to dietary intake if only slightly—so if you consume more saturated fatty acids or other fatty acids, you might contain more of them in your cell membranes, which in turn affects their functioning.[25]

Additionally, polyunsaturates in our membranes are used for another purpose—that is, to make the hormones called eicosanoids that operate inside our cells. In general, the omega-6s make pro-inflammatory eicosanoids, and the omega-3s make anti-inflammatory eicosanoids, which in turn influences our health.[26] So to some extent, if we eat more omega-6 or omega-3 polyunsaturates in our diet, we store more of either one in our membranes, which skews eicosanoid production. So, in short, how we consume our polyunsaturates, the omega-6s and -3s, determines our production of eicosanoids, which will influence our health.

FUNCTION: CATABOLISM

Our cells catabolize nearly all fatty acids to generate the energy molecule ATP. As noted before, our muscles and organs prefer fatty acids under normal levels of exertion, such as when we are sleeping, resting, walking or working. Or in other words, these tissues are catabolizing fats nearly all of the time. But as noted, our muscles prefer glucose at higher levels of exertion. Once inside our cells, fatty acids travel to our mitochondria, the powerhouse inside our cell, where they enter into the chemical pathway, called "fatty acid oxidation," whereby all types of fatty acids are transformed into one universal molecule, called acetyl-CoA, that then enters another chemical pathway called the Krebs cycle to produce ATP.

When fatty acids have longer tails, they produce more acetyl-CoA, and thus generate more ATP per molecule and smaller chains thus produce less ATP. The long-chain saturated fatty acid, palmitate, which is predominant in most animal fats and palm oil, generates around 120 molecules of ATP. The long-chain monounsaturated fatty acid, Oleic acid, abundant in most seed oils, especially olive oil, generates about the same. The medium-chain saturated fatty acid, lauric acid, that predominates in coconuts, generates about 78 molecules of ATP. The saturated short-chain fatty acid, acetic acid, generates about 12 ATP. Other short chains generate about half of that and as noted, we receive these fatty acids from fermented foods, including the ones in our colon.[27] In comparison, one molecule of glucose generates 36 ATP, as we saw in the prior chapter.

However, the longer chains produce ATP more slowly than medium and short chains, as well as glucose, because they require more time to transport and deconstruct into smaller molecules. Meanwhile, shorter chains, as well as glucose, produce ATP faster.[28] This explains why our muscles, in particular, switch from larger to smaller nutrients, like glucose, when needing energy faster for more intense activities. Additionally, per weight, rather than per molecule, all fatty acids produce about twice as many calories as glucose, with calories equating somewhat to ATP. So one gram of fat equals nine calories, while one gram of carbs equals four calories.

Generally, nutritionists say that one gram of protein equals four calories, but that is only after protein is converted into ketones and glucose, which, as noted, we usually do not want to happen. When under-consuming fatty acids, we pull them from our stores which lasts anywhere from days to months. But once we reach lower percentages of body fat, we start to slow our catabolism to guard against starvation.[29] Additionally, when under-consuming fats, we are prone to eating more carbs, especially refined ones, which can trend us towards the many detriments already mentioned.

When under-consuming the essential fatty acids, the polyunsaturates, we run the risk of several maladies and outright diseases — though, generally, this does not happen given the predominance of these fatty acids in most foods. When consuming the omega-6s and -3s in improper ratios, we also trend ourselves towards detriments.[30] When over-consuming fatty acids, we increase our risk for obesity as well as possibly various forms of disease, including heart disease. But we are way less prone to over-consume fats as compared to carbs for several reasons; additionally over-consuming fats is not as bad as over-eating carbs, especially refined carbs, for our health. When over-consuming saturated fatty acids, like palmitic and lauric acid, we possibly increase our risk of heart disease.[31]

SUMMARY

Again, from this short tutorial, you have already learned the basics of fats that you probably never heard before, which helps you optimize your diet.

- You can eat your rendered fats and oils either raw or cooked (premise 2).

- However, when fats are attached to other parts of our food, such as with nuts or meats, they are easier to digest when processed. You can soak, sprout, cook and grind nuts into butters. You can grind fatty meats into hamburgers and sausages or cook out the fat into broth and gravies (premise 2).

- Plant oils contain secondary compounds, so you should consume them with some caution even as some of those compounds, paradoxically, provide benefits (premise 3).

- You catabolize fats as your primary source of ATP for your muscles and most of your organs during rest and light to moderate levels of exercise (premise 6).

- Thus, when you exercise more at lower to moderate levels, you catabolize more fat. But during more intense forms of exercise, your muscles catabolize more glucose (premise 6).

- When fats have shorter chains, they make less ATP but faster. With longer chains, they make more ATP but slower.

- If you consume fats and carbs at the same time, your muscles and organs absorbs more of the glucose at first to lower your blood sugar but then returns to fats (premise 6).

- You anabolize fats into phospholipids for your cell membranes (premise 6).

- Our body stores fatty acids in our phospholipids in about these ratios: 30% saturated fat, 50% monounsaturated fat, and 20% polyunsaturated fat. This somewhat approximates the way we store them, too, in our phospholipids (premise 6).

- When you undereat fat, you merely catabolize more fat from your stores—which is efficient and helps you lose weight. However, once your body fat becomes lower, you may start to lower your metabolism or overeat carbs—both of which are detrimental (premise 6).

- When you overeat fat, you tend to gain weight, especially if you are eating moderate to high levels of carbs. However, when you eat more fat and fewer carbs, you paradoxically catabolize more fat—thus explaining why ketogenic diets do not cause weight gain and even help people lose weight (premise 6).

- Overeating fat does not seem nearly as detrimental as overeating carbs, unless you are overeating both of these at the same time—which trends you towards the diseases of civilization (premise 6).

- Polyunsaturates are essential fatty acids, meaning your body cannot make them so you must receive them in your diet to avoid detriments and even disease (premise 6).

- Polyunsaturates are in two forms: omega six and three. And you must balance them in your diet to avoid detriments (premise 6).

- Too many saturates, relative to monounsaturates and polyunsaturates, possibly affects your potential for heart disease (premise 6).

CHAPTER 5

VITAMINS, MINERALS, PHYTONUTRIENTS

I also cover vitamins, minerals and phytonutrients in this book—but only address them now in passing. Similar to essential amino acids and fatty acids, vitamins are also considered essential because our body cannot make them and thus we must receive them in our diet or suffer disease and eventually death. In almost all cases, these vitamins play roles in both catabolism and anabolism. However, in our natural and evolutionary diet, they are ubiquitous, so we did not need to make them. But in our modern diet, they are not necessarily ubiquitous enough because we can suffer deficiencies in some vitamins, perhaps even more than the experts believe, including in vitamin B12, A, especially preformed vitamin A, as well as vitamin C, D, E and K.

Macrominerals were already addressed as the animators of our bodies. While macronutrients, especially proteins and fats, create the structures of our body, the electrolytes, along with ATP, provide the spark within these structures that make us move, feel, think and dance. They potentially can become dysregulated and imbalanced, so I address them later.

Phytonutrients are not deemed essential; they are not even technically defined as nutrients, despite their name, because humans can clearly subsist without them or most of them. However, they probably provide enormous value to our health. They include plant pigments that give plants their color, like chlorophyll, carotenoids and several others. Some of these pigments lodge into our skin to protect us from the sun, into our brain to enhance our cognition, and the carotenoids convert into vitamin A. I also define organic acids as phytonutrients—that is, the compounds, like citric and malic acid, that give plants their tart and sour flavors. Other phytonutrients include flavonoids, lignans, phenols and many others. Generally, phytonutrients collectively fight microbes, thus enhancing our immunity, neutralize free radicals while providing multiple forms of benefits throughout the body to multiple systems.

CHAPTER 6
NUTRITIONAL PHYSIOLOGY SUMMARY

As you move forward into the other sections of this book, you will need to carry the knowledge from this section with you. For that reason, I summarize the relevant points below.

NUTRIENT CONVERSIONS

When not consuming certain nutrients in your diet, you can make most of them inside your body, so they are thus considered "nonessential" to your diet—and include many amino acids, fatty acids and even sugars in some senses. You can survive on about nine amino acids, two fatty acids, glucose, all vitamins and all minerals (even though you can recycle most minerals indefinitely), and you can make the rest inside your body. Your body can make the following conversions, and many more, mostly in your liver:

Type	Catagory	Name	Conversion A	Conversion B
Sugars		Glucose	Palmitate	
		Fructose	Glucose	
		Galactose	Glucose	
Fatty Acids	Saturated Fats	Lauric	Myristic	
		Palmitic	Stearate	
		Palmitic	Palmitoleic	
		Stearic	Oleic	
		Myristic	Myristoleic	
	Monounsaturated Fats	Palmitoleic	Oleic	
		Oleic	2 other monounsaturates	
	Polyunsaturated Fats	LA	Arachidonic	
		ALA	EPA and DHA	
Amino Acids	Glucogenic Only	Alanine	Glucose	
		Arginine	Glucose	Ornithine (urea cycle intermediate)*
		Asparagine	Glucose	Aspartate
		Aspartate	Glucose	Asparagine
		Cysteine	Glucose	
		Glutamate	Glucose	Glutamine, Proline, Arginine
		Glutamine	Glucose	Glutamate
		Glycine	Glucose	Serine
		Histidine	Glucose	Glutamate
		Methionine	Glucose	Cysteine
		Proline	Glucose	Glutamate

	Serine	Glucose	Glycine, Cysteine (with Met)
	Valine	Glucose	
	Threonine	Glucose	
Glucogenic & Ketogenic	Isoleucine	Glucose or ketones	
	Phenylalanine	Glucose or ketones	Tyrosine
	Tyrosine	Glucose or ketones	
	Tryptophan	Glucose or ketones	Alanine (partial breakdown)
	Threonine	Glucose or ketones	Glycine, Serine (less common)
Ketogenic Only	Leucine	Ketones	
	Lysine	Ketones	

TABLE 6.1- NUTRIENT CONVERSIONS

However, I obviously do not recommend that you only eat the non-essentials and then make all those conversions. Instead, I recommend that you consume all essential and non-essential nutrients, all nutrients in general, in their proper ratios to optimize your metabolism.

PROTEIN AND FAT: ANABOLISM

- You use protein for anabolism—that is, to maintain and grow your tissues. You should consume enough protein, with proper ratios of amino acids, to replace your discarded amino acids.

- You should strive to avoid using protein for catabolism.

- You should cook most of your proteins or process them in other ways. As compared to plant proteins, animal proteins are generally more digestable and complete. However, you can combine and prepare plant proteins to make them possibly as complete and digestible.

- If you under-consume amino acids, you then pull them from your muscles. If you over-consume amino acids, you then just convert them to glucose and ketones, which you do not need.

- If you consume your amino acids in the wrong ratios, then you also generate all the detriments listed above and even more of them.

- You generally want to consume your amino acids in ratios that match both complete and collagen proteins.

- All humans anabolize protein in somewhat equal amounts per our lean mass (as fat mass does not require much protein). However, we slightly increase our need for protein as we increase the duration and intensity of our exercise.

- You use small amounts of fats to anabolize the phospholipids in your cell membranes—but for that purpose, you only use the long chains and about equal amounts of saturates, monounsaturates and polyunsaturates.

CARBS AND FAT: CATABOLISM

- You use large amounts of carbohydrates and fat for catabolism—for the production of ATP.

- Our brain and red blood cells prefer glucose, while our brain also catabolizes other nutrients in small amounts, under certain conditions, including lactate, acetate, and ketones.

- When you under-consume carbs, you then convert other nutrients into glucose or ketones.

- When you over-consume carbs, especially without fiber, you potentially cause all sorts of detriments for yourself, including glycation, oxidation, lipogenesis and hormonal imbalances.

  In fact, it appears, based on various lines of evidence, that this contributes more than anything to the diseases of civilization.

- You cannot boost the metabolism of your brain by consuming more carbs than necessary, because the brain maintains the same level of metabolism under nearly all conditions.

- Fructose and galactose are not metabolized by the body, so they are converted into glucose; they also have other detriments. For that reason, you should consume them in moderation.

- Fiber helps your body in many ways. It helps regulate the flow of food through your gut as well as the absorption of nutrients. Once reaching the colon, soluble fiber, in particular, provides an enormous array of additional nutrients, including vitamins, minerals, phytonutrients, antioxidants, especially hydrogen gas, as well as additional catabolic nutrients such as the short chain saturated fatty acids, including propionic acid, butyric acid, which is particularly beneficial for the colon itself, and acetic acid which provides multiple benefits. When over-consuming fiber, we risk filling ourselves too much and thus not eating enough macronutrients, which then might lead to malnutrition.

- Our muscles and most of our organs prefer to catabolize fatty acids under normal levels of exertion and then switch to glucse during higher levels of exertion.

- When you under-consume fats, you draw from stores that, in most people, can last for weeks and even months. However, at some point, you risk slowing your catabolism and overeating carbs and the problems that come with that.

- When you over-consume fats, you risk hyperlipidemia, obesity, possibly insulin resistance and other conditions, especially when you consume at the same time moderate to high levels of carbs.

  However, overeating fats is not as addictive or detrimental as overeating carbs — and even, possibly, provides benefits under certain conditions.

- When you over-consume some forms of saturated fat, you may trend towards heart disease.

- If you under-consume polyunsaturates, you trend towards disease. If you over-consume them, you likely risk other detriments.

- When you consume omega-6 and -3 in the wrong ratios, you trend towards disease.

- Generally, all humans catabolize about the same number of calories to maintain enough heat in their bodies. However, they obviously can consume more or fewer fats relative to carbs for those calories.

- However, we increase our need for calories when we exercise. If our exercise is more intense, we need more carbs relative to fats, because our muscles catabolize glucose at higher levels of exertion. If less intense, you need more fat relative to carbs.

- Longer chain fatty acids provide more ATP but do so more slowly. Shorter chains produce less ATP but more quickly, while glucose is in between. Also, per weight, fat contains about twice as many calories as sugar, with calories equating, roughly, to ATP.

- Generally, when cells metabolize faster, they prefer smaller nutrients that catabolize faster; thus the brain metabolizes faster and prefers glucose and smaller molecules like ketones, acetate, and lactate. Meanwhile, our muscles, at low to moderate exertions, metabolize more slowly and thus prefer longer chain fatty acids for more sustained energy, but then switch to glucose at higher levels of exertion and then lactate.

Type	Nutrients	Amount of ATP	Rate	Calories/g	Note
Fatty Acid	Acetic acid	10 ATP	Fast	9 cal	colonic fat
	Propionic acid	18 ATP	Fast	9 cal	colonic fat
	Butyric acid	27 ATP	Medium	9 cal	colonic fat
	Lauric acid (C12)	80 ATP	Slow	9 cal	dietary fat
	Myristic acid (C14)	94 ATP	Slow	9 cal	dietary fat
	Palmitic acid (C16)	106 ATP	Slow	9 cal	dietary fat
	Stearic acid (C18)	120 ATP	Fast	9 cal	dietary fat
	Oleic acid	118 ATP	Fast	9 cal	dietary fat
	Palmitoleic acid	104 ATP	Fast	9 cal	dietary fat
	Lineoleic acid	110 ATP	Slow	9 cal	dietary fat
	Alphalineoleic acid	109 ATP	Slow	9 cal	dietary fat
	Arachidonic acid (AA)	122 ATP	Slow	9 cal	dietary fat
	Eicosapentaenoic acid (EPA)	120 ATP	Slow	9 cal	dietary fat
	Docosahexaenoic acid (DHA)	130 ATP	Slow	9 cal	dietary fat
Sugars	Glucose	30–32 ATP	Medium	4 cal	
Misc	Ethanol (alcohol)	12 ATP	Fast	7 cal	
	Lactic acid	15–17 ATP	Fast	< 1 cal	fermented food
	Ketones (-OHB)	22 ATP	Medium	2 cal	endogenous
Protein	Amino acids on average	~20 ATP	Slow	4 cal	only after conversion

TABLE 6.2- AMOUNT OF ATP AND CALORIES PER MOL-
ECULE OF NUTRIENTS.

CONCLUSION

So in short, you need to consume all nutrients in their proper ratios to optimize your metabolism, which in turn greatly reduces your probability of various maladies and diseases. Eat the right amounts and ratios of protein for anabolism of all tissues. Eat the right amount of carbs with their fiber intact for catabolism for your brain and red blood cells, and eat considerably more glucose than fructose or galactose. Eat the right amounts and ratios of fats for your muscles and organs, making sure you consume enough of the essentials first—that is, the polyunsaturates in the proper ratios. Then determine your

ratio of monounsaturates to saturates. If you increase your level of exercise, eat more calories overall and slightly more protein in particular. With intense exercise, eat more carbs; less intense exercise, eat more fat.

In the next section on ancestral consumption, I will explain how our ancestors evolved to do all of the above, in various and sometimes imperfect ways, even as they encountered considerable obstacles. In the following section, on premises, I will further conclude on this point and then clarify how you can, fairly easily, balance your nutrients in our modern world to optimize your metabolism and avoid various maladies and diseases.

PART 2

ANCESTAL CONSUMPTION

To understand our optimum diet, why do we need to understand ancestral consumption? Why do we need to understand the diet of our ancestors? The answer is because evolution designs all organisms through natural selection to eat specific diets that help them survive, thrive and reproduce—essentially to optimize their biological potential. Evolution designs organisms to acquire, ingest and digest certain foods and then metabolize, store and convert the underlying nutrients. When eating those foods, organisms are then in alignment with their design—giving them optimum health. As evidence of this, merely watch videos of any animals in their natural habitat, including humans. Do they look diseased? Weak? Overweight? Or permanently stressed? Or unable to deal with the challenges of their life? Instead, in their unique and specific ways, they usually look radiant, powerful, in perfect health, capable of performing incredible feats and perfectly adapted to their environment.

So should we not understand our ancestral diet to understand our own optimum diet? And given that we have not changed all that much over thousands of years, should we not, too, question whether any new or modern diets are better than our ancestral diet? In the pages ahead, I will describe how we adapted to our diet over billions and, more specifically, the millions of years of the human lineage, and reveal the specifics of that diet along the way.

IMAGE 7.1- SQUIRREL

CHAPTER 7

PROBLEMS WITH THE PALEO DIET

Many people are familiar with the concept of "ancestral consumption" through another concept called "The Paleolithic Diet," which is mostly explained in this way:

We humans consumed Paleo diets for 300 thousand years, and really much longer than that, eating diets high in meat but lower in fats and carbs, along with many plant foods like nuts, fruits, and vegetables. However, we only consumed our agricultural diets, which mostly consist of grains and legumes, over the past ten thousand years. Since unable to evolve adaptations to this agricultural diet within ten thousand years, humans therefore evolved for paleo diets—so we moderns must eat Paleo to assure our health.

While truthful in some ways, this concept is wrong in others. First of all, our ancestors never consumed any ubiquitous Paleo diet but many different types of diets with different ratios of nutrients, as I will prove in the pages ahead. Secondly, and oddly enough, our ancestors never consistently consumed the diet above either because humans cannot tolerate diets full of lean meat with few carbs and fats for long. Thirdly, we moderns do not necessarily have the same genes as these ancestors because we are separated from them in most cases by thousands of years: while nearly all of our genes evolved prior to ten thousand years ago, we nonetheless continued to evolve some genes through agriculture all the way to the present, and adapted to new foods.

Besides, we do not live the same lifestyle as Paleolithic people, so should we question whether we should eat their diet? Paleo people exercised way more than us all day long, walking, jogging, carrying, kneeling, climbing, etc. Even when resting they did not sit in chairs, which disengages their muscles, but squatted, which engaged their muscles.[32]

IMAGE 7.2- FORAGER, SQUATTING

Additionally, they frequently inhabited cold environments, requiring them to generate more metabolism to keep themselves warm.[33] For this reason, as compared to us, they needed more calories for their greater metabolism, and they needed more of those calories from protein and fat for low to moderate levels of exercise. However, they did not need any more carbs than us—because humans only need more carbs at higher levels of exertion, such as sprinting, which was not all that common for foragers.

While the concepts behind the Paleo Diet are flawed, the concept of ancestral consumption is not, when presented with greater truth and accuracy. In the pages ahead, I will explain the framework that

I use to describe ancestral consumption—that is, the foodways equation. After that, I then tell the actual story of ancestral consumption, starting from the beginning with the emergence of single cells, then animals and primates. But I mostly focus on hominoids, starting with our theoretical ancestor that lived millions of years ago. I extend the story through *Australopithecus*, *Homo erectus* and, in the end, *Homo sapiens* through three different eras: as foragers, agriculturalists, and moderns. In my other books, the series known as "The First Supper," I write about the evolution of every part of ancestral consumption systematically and with great detail and precise footnoting. In this particular version, I am presenting the summarized and generalized version, which, I surmise, most specialists in this field would think reasonable to some degree or another.

FOODWAYS EQUATIONS

As you probably remember from high school biology, all animals, including humans, are born with four basic needs for survival: protection from their environment; avoidance of their predators; reproduction; and lastly, and most importantly, the acquisition of foods—otherwise known in this book as foodways equation. On one side of this foodways equation are the calories and macronutrients the animal consumes. On the other side of the equation is the metabolism necessary to acquire that food and utilize the nutrients. To acquire food in their environment, most animals follow this basic chronology. They sense food in their environment, move (locomate) to that food, capture that food through hunting and gathering, and then ingest that food with their mouths and teeth. Once the food is digested into nutrients, those nutrients then undergo conversion, storage and then metabolism, either catabolism or anabolism. Or in short on one side of the equation is calories; and on the other side of the equation is metabolism needed to acquire and utilize those calories. This is somewhat similar to our idea of our calorie intake should equal our metabolism (or calorie burn) so we stay in balance.

Animals also use their brains to both acquire and utilize their food. They use their autonomic and somatic nervous system to sense and capture their food. But they also use their cognitive brain to identify, qualify, and quantify that food, and strategize ways to acquire that food while overcoming problems. They may also use their cognition to cooperate with their conspecifics to acquire their food.

Interestingly, these foodways metabolisms require nearly every tissue in our body, including our eyes, bones, muscles, mouth, teeth, jaws, gut, pancreas, gallbladder, liver and kidney, as well as our heart to pump nutrients to our cells and our lungs to deliver oxygen for catabolism. So all animals, including humans in their original condition, are in fact almost entirely designed for their foodways. As seen from the perspective of biology, we are really eating machines. In simplest terms, in the wild world, we eat calories so we can burn calories to attain calories. In the modern world, we eat calories and then force ourselves to exercise to burn those calories.

OBSTACLES AND DEFENSES

For animals in the wild, including humans, the foodways equation is more complicated because animals encounter enormous numbers of obstacles and defenses when acquiring their food. First, their food is typically both scarce and scattered across their territories, while only appearing at certain times of the year in many cases. Fruit trees are scattered across jungles, for example, and only bear fruit at certain times of the year. Secondly, they must compete with other animals and even microorganisms that eat the same food. And finally, their food is defended. For example, animal foods are hard to catch as they run, fly away, or fight back, and plant foods, as we have seen, contain mechanical and chemical defenses called antinutrients and toxins.

So animals must overcome all those obstacles and defenses, which in the end pushes them to evolve higher levels of efficiency and competency in their foodways. They not only must sense their food, but sense that food, in some cases, from miles away with incredible eyes and noses. They not only must locomote, but locomote with great efficiency and speed to travel long distances and arrive at their food before their competitors. They locomote like swallows, cheetahs, and mahi mahi—and Roger Federer. They not only must capture their food, but they do so in incredible and precise ways.

Although the foodways equations seem distant to us humans, our primate and hominoid ancestors lived within this foodways equation every day of their lives while encountering all the obstacles and defenses. To understand this point, imagine yourself lost in the wilderness with several of your friends. You suddenly rediscover your senses, especially your eyes, to help you locate your food, scanning for animals or colorful plants. You hear every nuance in your surroundings. You walk seven miles per day, climbing trees, fording rivers and creeks, crawling under brambles and bushes. You use your hands and fingers to capture animals, to pick berries, and you build tools for those same purposes. You are much more aware of your body as you ingest, digest and metabolize your food, feeling the energy return to your body. You use your brain to identify, count and qualify your food—and to form strategies and solve problems. You cooperate, compete and communicate with your friends for that food. Along the way, you encounter incredible obstacles and defenses around that food when animals move away or fight back, or plants coat themselves in thick, fibrous husks and occasionally poison you with their toxins, even as other animals, even possibly your fellow humans, try to steal your own food. In the back of your mind, every day, you are calculating your foodways equation: can I acquire enough nutrients to match the metabolism I use to acquire those nutrients, and then, hopefully, some excess as well. While feeling totally challenged and even afraid, you might also feel for the first time most like yourself, and most fully alive, as you fulfill your primal and evolutionary charge.

CHAPTER 8

ADAPTATIONS

FOODWAY ADAPTATIONS

Generally, when animals inhabit stable environments and ecosystems, they are already adapted to keep their foodways equation in balance; they consume about the same number of calories as they metabolize, even while overcoming all of their obstacles and defenses. But sometimes their ecosystem changes, making their food either more or less available—thus unbalancing the equation. If acquiring more calories than needed, they breed more, so there is less food per animal—and thus rebalance the equation. If acquiring fewer calories than needed, they breed less, which then creates more food per animal—thus rebalancing the equation.

However, in some cases animals still cannot acquire enough food because their ecosystem changes substantially, due to alterations in geography, weather, flora, fauna, and other factors. This, in turn, causes their food supply to diminish or change. And in that case, animals are not adapted to their environment anymore, so they must evolve or risk perishing as individuals or groups.[34] They adapt by using four different methods that usually happen in this order: inherent flexibility, epigenetics, cultural evolution, and, lastly, natural and sexual selection.

INHERENT FLEXIBILITY

Through "inherent flexibility," animals adapt to temporary changes in their environment by using traits already present in their genes or intelligence. Usually this means they switch from preferred foods to fallback foods.[35] For example, many primates subsist mostly on fruits during most of the year, but sometimes fruit becomes scarce due to various causes, so they adapt by eating other foods like leaves, piths, or tubers.

These primates already possess the traits necessary to acquire those foods and utilize their nutrients. In some senses, fallback foods are just as important as preferred ones, because without them the primates could not survive.

EPIGENETICS

However, in many cases, animals cannot adapt through "inherent flexibilities," so they may rely on "epigenetics." In this cases, when animals encounter environmental changes, they adapt by forming compounds that either turn genes on or off, and to various degress — thus changing the traits of the animal. Sometimes animals then pass these compounds down to their offspring. In humans scientists know that some foods will trigger the expression of certain genes, and some foods, especially bad ones, trigger genes for disease.[36] Animals can adapt in this way assumably over shorter periods of time.

However, scientists thus far do not understand epigenetics all that well — but the effects likely range from minimal to considerable.[37]

CULTURAL EVOLUTION

In many cases, animals cannot adapt through "inherent flexibilities" or "epigenetics," so at that point, some animals with larger brains then resort to "cultural evolution." They use their intelligence and creativity to adapt behaviors, otherwise known as memes, that are then handed down to their offspring and thus perpetuated, similar to genes. Although most animals are not capable of "cultural evolution" due to limited intelligence, many primates are capable of this form of adaptation, usually through the use of tools. For example, many primates are able to crack open nuts with rocks, collect ants, termites and honey with sticks, and then teach their offspring these same methods. Some birds, octopuses, and cephalopods exhibit similar behaviors for their foodways.[38]

Once reaching certain levels of intelligence, about the time of Homo erectus, our own direct ancestors used cultural evolution as their primary form of adaptation to changing and new environments. They invented tools for acquiring foods, fire for detoxifying and cooking foods, strategies for hunting animals, etc., which in turn allowed them to inhabit nearly every ecosystem on earth. Cultural evolution can happen potentially within moments, and it does not require any changes to morphology.

NATURAL AND SEXUAL SELECTION

Finally, in some cases, unable to adapt through "inherent flexibility," "epigenetics," or "cultural evolution," animals must adapt through the most famous of all adaptation mechanisms: "natural selection." In this case, animals change their genes, which then changes the expression of their morphology or functioning, which then readapts them to their environment. This, in turn, allows them to gather more food, which in turn helps them reproduce more, thereby pushing those genes into future generations.

Even when adapted to stable ecosystems, animals are changing their genes all the time through mostly accidental and random causes such as mutations, damage, or drift, as well as through reproduction. But in this case the animals do not need these genes because they are already well-adapted, so these genes just fade away. But when the environment is unstable, some of these genes, perhaps just several out of millions, can help animals better adapt to acquire foods, to attracts mates, and then propagates those genes into the future.

Unlike the other forms of adaptation, natural selection takes thousands and even millions of years to occur, while many animals die in the process. So, while perhaps the most effective form of adaptation, natural selection is the slowest and, in some senses, most painful.

Animals also adapt their foodways, possibly, through sexual selection, where animals, including humans, sometimes reproduce because of the sexual traits of the mate—that is, traits that make them "sexier" but not more well adapted. Scientists propose many different theories for this, but the one that seems to make the most sense is that those traits advertise success in foodways. To use one of the classic examples, female peacocks prefer male peacocks with greater plumage even though those feathers do not even help the male fly, escape predators or gather foods while requiring extra nutrients. However, that same plumage possibly advertises to the female that the male was so successful in their foodways that they acquired additional nutrients to metabolize those useless feathers. We can easily think of all the interesting and funny ways that humans display sexual selection—the things we find sexy that do not help us survive.

SUMMARY

All animals, including humans, exist within foodways equations. On one side of the equation is calories; on the other is the metabolism—the steps needed to acquire food and utilize its nutrients. However, while acquiring their food, animals encounter obstacles and defense along the way—that is, their food is distant, scattered, defended and under competition from other organisms. But to overcome these obstacles, animals do not just need foodways but superlative foodways—thus explaining why all animals are magnificent in their design.

Most animals are well adapted in their foodways equation when their environment is stable. However, when their environment changes, their foodways equation can be imbalanced, meaning they cannot acquire enough food or the right kind of foods. In that case, they reduce reproduction so there are fewer animals per the available food. But in some cases, especially when the ecosystem changes substantially, along with the available food, they still cannot acquire enough food, which threatens their existence. And in this case, they must adapt through inherent flexibility, epigenetics, cultural evolution, and natural and sexual selection.

Our ancestors evolved through this foodways equation. They adapted to changing environment — through all four forms of adaptation while overcoming all those obstacles. On one side of the equation, they increased their calories and on the other side, their metabolism, due mostly to larger bodies and bigger brains; they also decreased their metabolism in some ways. Thus, through understanding this equation, we can explain their evolution. Below is the schematic of the foodways equation — which I use in the pages ahead to explain the further evolution of our ancestors. While these equations are good at representing our evolution, they are extremely abbreviated and thus are not mathematically accurate in this presentation.

FOODWAYS EQUATION

CALORIES

=

MACRONUTRIENTS

=

METABOLISM

=

Sensing:

+

Locomotion:

+

Capture: (hunting and gathering):

+

Ingestion:

+

Digestion:

+

Liver: (conversions, storage and detoxification):

+

Brain: (cognition and sociality):

CHAPTER 9

PROKARYOTES, EUKARYOTES, ANIMALS, AND PRIMATES

———

In the prior section, as applied to all animals, I summarized the food-ways equation, as well as obstacles and adaptations to that equation. In this section I will apply all of that to our own most immediate ancestors. I begin the story with the formation of nutrients and the first protocells and cells billions of years ago. I then quickly cover animals and mammals in general before focusing on primates and then three hominoids in particular. I then focus on *Homo sapiens* through three eras: hunter-gatherers or foragers, which is the term I will use going forward; agriculturalists and then moderns, which refers to ourselves over the past couple of hundred years. In precise terminology, we humans are primates, apes, hominoids, and of the genus *Homo*. However, for the sake of simplicity and clarity, I generally use the term primates to refer to monkeys and non-human apes, hominoids to refer to non-human hominoids (*Australopithecus* and *Homo erectus*), and I refer to ourselves as *Homo sapiens* or humans.

Four basic elements or atoms create nearly all of life: carbon, oxygen, hydrogen, and nitrogen. These atoms are abundant in the Earth's crust and, for various reasons, are prone to combine with each other in various configurations. Back billions of years ago, through processes that are not well understood, these atoms combined to generate the macronutrients: sugars, amino acids and fatty acids.

CELLS

Also in ways that are not understood, these macronutrients formed into the first units of life: cells—in this case, prokaryotes, otherwise known as bacteria and archaea. Since the atmosphere contained little oxygen, these prokaryotes were anaerobic, meaning they could neither tolerate nor utilize atmospheric oxygen. Some billions of years later, many of these prokaryotes became photosynthetic, making glucose from sunlight while releasing oxygen into the atmosphere. Paradoxically, that oxygen, which is inherently unstable, was toxic to them, so their own existence was threatened. But through natural selection, they evolved. Some prokaryotes remained anaerobic but slunk away into places with low oxygen, and these bacteria still inhabit places like our colon and our fermentation jars. Others became aerobic, and they evolved antioxidants to combat the damages of oxygen.

Some also evolved to utilize that oxygen through another, more complex form of catabolism, called aerobic respiration, which used the energy inherent in oxygen to generate even more ATP, leading essentially to more powerful cells. Sometime thereafter, larger bacteria engulfed smaller, aerobic bacteria, and in time, these smaller bacteria evolved into the mitochondria of the larger bacteria. This, in turn, led to the evolution of the second units of life: eukaryotes, single cells, that were much larger, more complex and more powerful than prokaryotes.

Some of these eukaryotes were like tiny animals with all the same foodways as ourselves. They sensed their food with primitive eyes; locomoted to their food with tiny legs; ingested their food with mouths and digested that food into nutrients. They then utilized those nutrients for metabolism or converted or stored them. They also all contained the same basic nutritional physiology as us. For example, they catabolized glucose and other molecules to generate ATP, and they anabolized amino acids into protein for their bodies and fatty acids into phospholipids. So even back billions of years ago, evolution had already designed the microscopic blueprint for animals and humans. Even now, we are humans are amalgamations of eukaryotes that are similar to the ones from billions of years ago. Generally, evolution works this way—that is, once establishing something that works, it tends then to replicate that pattern over again for billions of years.

ANIMALS

Soon after evolving, eukaryotes bonded with other eukaryotes to create multicelled organisms that were just blobs in the beginning. But soon those cells started to differentiate to perform different types of work; some cells became sensors, mouths, feet, stomachs and etc. Once more fully evolved in this way, all animals established the usual foodways but adapted them to their specific ecosystem. All sensed food in their environment, but some relied more on sight, others on smell, others on sound. They all locomoted to their food but some walked, others flew, and others swam. They all ingested their foods but contained different beaks, snouts and mouths. They all digested their food, but evolved different guts for different foods. However, once digesting their food into nutrients, most animals, especially mammals, appear to transport, store, convert and metabolize and anabolize nutrients in similar ways.

PRIMATES

Jumping ahead to around 250 million years ago, the reptiles, known as dinosaurs, dominated most of the Earth while the mammals scurried around under their feet. But about sixty million years ago, the comet known as Chicxulub hit the Earth, sending most of the dinosaurs into extinction while smaller reptiles survived. Meanwhile, the mammals also survived and even thrived, becoming the largest and most dominant species on earth, filling nearly every ecosystem. During this time one of these mammals evolved into the first known primates—an animal, about the size of the current shrew, called the Teilhardina, that specialized on eating insects.

IMAGE 9.1- TEILHARDINA

However, over the next sixty million years, this primate evolved into many different primates that radiated all around the world, with nearly all of them inhabiting various jungles and niches within those jungles. Some evolved into folivores, eating mostly leaves; others into frugivores, eating mostly fruit; some into insectivores. And some evolved, specifically, in Africa about thirty million years ago as omnivores, specializing in both fruits as well as small amounts of animal foods like insects and arthropods. And about seven million years ago, one of these omnivores evolved into what I am calling our "theoretical ancestor," that is, the ape that evolved both into Chimpanzees as well as our own line of evolution.[39]

CHAPTER 10

THE COMMON THEORETICAL ANCESTOR

Scientists have not yet found any fossils of this "theoretical ancestor" that launched our line of evolution. So they speculate about him based on their understanding of our evolution and our genetics. For the purposes of this book, I suggest that this ancestor was likely similar to extant chimpanzees, and especially Savannah Chimps, for several reasons. First, they likely lived in similar ecosystems as current Chimpanzees—that is, either dense or scattered jungles—because that is about all that existed back then in Africa. Secondly, their brains and bodies were probably about the same size as Chimps, because they evolved into Chimps as well as *Australopithecus* who was also about that same size. And thirdly, chimps and bonobos are more genetically similar to humans than any other species, suggesting that our ancestor was also similar to them. Thus, in this next section I will essentially reconstruct the foodways of the "theoretical ancestor" through determining the foodways of the Chimpanzee—and thus use the "common ancestor" and chimps synonymously—even while acknowledging that this exercise is somewhat theoretical.

IMAGE 10.1- CHIMP

Our "theoretical ancestor" inhabited tropical rainforests, rich in trees and foliage, because that is all that existed in Africa back then. They were on average four feet tall and weighed 100 pounds, while probably possessing the most encephalized brain of their time, although only one-third the size of our own. They lived in small troops of around 120 individuals, while sleeping in trees for protection. During the day, they radiated outward in their territories in smaller troops, using various forms of cooperation, including communication, to find and capture their food. In their division of labor, males gathered plants, hunted occasionally, and patrolled their borders to protect their territories from neighboring chimps while one alpha, along with his allies, governed the group. Females also gathered plants, nursed their children for three years without ever leaving their side, and also engaged in the governance of the group with one alpha and her allies. The males were patrilocal, meaning they tended to stay in the troop of their origin, allowing them to better bond with each other and thus govern the troop; meanwhile the females left for other troops. The sexual dimorphism was minimal, meaning the males were not much bigger than the females. In all primates, when males are much bigger than females, the males tend to dominate the females; but when they are closer in size, they tend to cooperate more with each other. So relative to many other primates, male and female chimps were more cooperative.

In some cases, they shared food with each other—that is, mothers shared with their children, and males shared animal flesh with females, sometimes in exchange for sex. However, they were not cooperative with neighboring chimps since they were competing for food, territory, and mates and, at times, engaged in combat with their neighbors, sometimes violently, which in some cases included rape and infanticide.

Does this sound familiar?

Interestingly, we humans in general exhibit most of these same characteristics, although with variations, and presumably most of our line of evolution likely did as well. For this reason, and for the sake of brevity, I do not discuss social structure and lifestyle much again until arriving at agricultural and then modern humans, when all this changed dramatically.

ROUGH AND REFINED FOODS

While many primates during that time consumed rough foods, our ancestor consumed refined foods. As you might imagine, rough foods include mature leaves, grasses, stems, twigs, bark, roots and the like, which are higher in fiber, antinutrients and toxins and lower in macronutrients. These foods contain fewer obstacles to attain; they are neither scarce, scattered nor mobile. They are generally everywhere and stationary, so for primates they are easy to acquire and generally not under much competition. Interestingly, certain primates, usually known as folivores (leaf eaters), evolved foodways for these types of foods. On one side of the equation, these folivores essentially consume fewer calories, and on the other side, show less metabolism, most especially in their sensing, locomotion, and capture, as well as their brain. They just do not need to move very far, very fast, or with great dexterity, to pick some leaves or pull some grass; at the same time, they do not need to be all that smart to acquire these foods. However, they need more metabolism for ingestion and digestion to break down these foods and extract nutrients from fiber, and more metabolism for their liver to detoxify plants and convert nutrients.

Refined foods are lower in fiber, antinutrients, and toxins as well as higher in macronutrients. These foods include fruits, tubers, grains, legumes, as well as animals and possibly some leaves like young, tender leaves. These plants contain more obstacles as they are more scattered in time and place, harder to gather and under more competition. Fruits are found only on some trees and only at certain times of the year, and then usually under considerable demand from multiple species. Tubers are under the ground, and grains are scattered and tiny and difficult to process. Prey animals are usually more scattered in their territory and able to move away, hide, fight back, etc. So, these foods are harder for primates to acquire, requiring more sensing, locomotion, acquisition and brains.

Certain primates evolved foodways to acquire these refined foods. On one side of the equation, they acquire more calories and macronutrients from these refined foods, and on the other side, they exercise more metabolism for sensing, especially vision, for locomotion to cover greater and more complex terrain, and for acquisition for more complex gathering and hunting. And they have higher metabolic needs for their larger brains, which in turn helps them to acquire more complex foods, and in some cases, work with their conspecifics to acquire those foods. However, they require less metabolism for ingestion, digestion, and conversion because their foods are easier to chew, digest and require less conversion and detoxification.

These primates are defined as omnivores and we of course evolved from them.

As compared to folivores and most other mammals, our theoretical ancestor consumed more refined diets. More specifically, they consumed tree foods with fruits at over 80% of their diet by weight; leaves about 10% and animals about 5%, including ants, termites and other insects, while the males periodically hunted monkeys. On average they consumed about 1900 calories per day.[40] Based on various lines of reasoning and evidence, I predict they consumed about 20% of their calories from protein (about 48 g per day),[41] 60% from carbs (about 280 g per day), about 10% from dietary fat (about 20 g per day) and 10% from colonic fat—that is, the short-chain fatty acids from eating over 100 g of fiber per day with 50 g of that fermentable.[42]

CATABOLISM

As we know, animal brains catabolize glucose, acetate or ketones, and our common ancestor, with their larger brains, consumed fruit which provided both glucose directly, acetate (aka acetic acid), and propionate (which is converted into glucose) from the fermentation of fiber in that fruit.[43] Fruits are also rich in potassium, phytonutrients, especially the organic acids, as well the vitamins A, B, C, and K. Of all the plants, especially the ones providing carbs, fruits are lowest in antinutrients and toxins: as explained before, plants want to encourage predation of their fruits. Animal muscles and organs prefer to catabolize dietary and colonic fatty acids. So, our common ancestors used these nutrients for these tissues, while also converting excess carbs into these nutrients and then used the glucose in particular more directly.

ANABOLISM

Our common ancestor consumed most of their proteins from plants, specifically leaves, but also from small amounts of animals. As noted before, plants contain lower quality proteins because they contain more fiber, antinutrients, and toxins, as well as imbalanced ratios of amino acids. But they also consumed small amounts of whole animals, including the bones, muscles, tendons, tongue, adipose, organs, brain and heart, while tending to prioritize fat and liver. So, our theoretical ancestor consumed lower-quality protein from plants combined with higher-quality proteins from whole animals that contained balanced ratios of complete and collagen protein, as well as plenty of vitamins and preformed vitamin A.

They also received their fat from both plants and animals—but not much because jungle plants and animals tend not to carry much fat. Altogether, they probably consumed about 20% saturated, 30% monounsaturated and 50% as polyunsaturated fat, at probably two to four parts omega-6 to one part omega-3, both the shorter and longer chain versions.

FOODWAYS EQUATIONS

With the common ancestor, we already see the relationships in foodways that defined the evolution of the remainder of our ancestors, leading all the way to foraging humans. On one side of the equation, diet (food and nutrients), our common ancestor already consumed more refined foods: sweeter fruits, tender leaves, some seeds and nuts, and smaller amounts of animal foods, mostly insects but also other primates. This diet was lower in fiber (though still plentiful in fiber), toxins, and antinutrients, and richer in macronutrients in somewhat proper ratios to each other, to optimize anabolism and catabolism.

On the other side of the equation, metabolism, our ancestors possessed all of the best kinds of traits to help them acquire this refined diet. They sensed their food mostly with their eyes, with superior color and stereoscopic vision—all of which was perfect for identifying fruits and insects in complex and layered jungles. We humans possess mostly the same eyesight. They locomoted with arms and legs, perfect for climbing trees and accessing fruits and leaves, and even chasing down prey while able to knucklewalk or walk upright for short distances from one tree to another. With opposable thumbs and nimble fingers, they gathered their fruits and leaves from various angles; with great speed and strength, they hunted other animals, especially monkeys. They ingested their fruits and leaves with jaws and teeth perfect for their diet, canines for shredding fruits and leaves, and then molars for grinding those foods into pulps, chewing on average about five to six hours every day.[44] They digested their food with guts perfect for their refined diet—that is, relatively small guts similar to our own, with acidic stomachs but small intestines and colons for fermenting fiber.

They had larger brains, too, which in turn allowed them to overcome more of the obstacles to finding their food. These brains allowed them to conceptualize larger terrains for more scattered foods, remember when foods were available in those terrains, identify, quantify, and qualify many types of foods, solve problems to acquire that food and, perhaps most of all, cooperate strategically with their conspecifics to acquire that food while competing with other animals and rival chimps.

Once digesting and absorbing their food, they transported, metabolized, stored and converted nutrients in their body similar to other mammals, including ourselves. But they differed from us in some ways, too: they stored less fat than ourselves on their bodies and performed more conversions overall. They converted enormous amounts of fructose from fruit into glucose; glucose into palmitate; and some amino acids into other amino acids.

Their foodways equation looked like this, as compared to folivores, especially:

FOODWAYS EQUATION FOR THE THEORETICAL ANCESTOR AS COMPARED TO FOLIVOROUS PRIMATES

CALORIES: 1900 per day.

=

MACRONUTRIENTS: from more refined diets with less fiber, toxins, anitnutrients and more macronutrients.

=

Protein: 20% of calories.
Completes: low from eating more plants than animals.
Collagen: low from eating more plants than animals.

+

Carbs: 60% of calories:
Fruits: 55%
Starches: 5%

+

Fats: 20% of calories:
Dietary: 10%
Colonic: 10%

=

METABOLISM

Sensing: eyesight increased over nearly all mammals to help them locate harder-to-find foods.

+

Locomotion: increased over folivores to help them cover more ground faster for more scattered and harder to acquire food.

+

Capture: (hunting and gathering): increased for the same reason as above.

+

Ingestion: decreased over folivores because refined foods need less chewing.

+

Digestion: decreased because refined foods need less digestion.

+

Liver: (conversions, storage and detoxification): decreased over folivores due to more balanced nutrients and fewer toxins in fruits and animals.

+

Brain: (cognition and sociality): increased over folivores substantially because larger brains are better at acquiring more refined diets which in turn provided better nutrients for the brain.

As we shall see, these dynamics drove our evolution from the theoretical ancestor (and probably well before) all the way to foraging humans—that is, for about five million years or so. They increased calories; and increased and decreased metabolism; but in the end their foodways equation stayed somewhat balanced.

CHAPTER 11
AUSTRALOPITHICUS

EVIDENCE

Since Australopithecus and other hominoids are extinct, scientists cannot observe them directly, nor use any other animal as their model. For this reason, we understand their foodways only indirectly, using various methods. First, scientists study their fossils, usually just fragments of skulls, teeth, and bones and from this extrapolate the overall shape and length of their body as well as the size of their jaw, teeth, skull and thus their brain. They study the morphology of their teeth to determine the diet they were designed to process;[46] they also study the fossilized food lodged into those teeth,[47] as well as the wear on those teeth.[48]

They also study the carbon in the enamel in their teeth. During photosynthesis, plants produce either carbon three or four, and when animals eat those plants, or animals that eat those plants, those types of carbons lodge in in their enamel. And with this information, scientists can predict their diet.

They also study fossilized coprolites, basically petrified crap, which holds the residues of food.[50] They also reconstruct the environment of Australopithecus by considering both fossilized plants and other animals that date around that same time or, otherwise, are found in the same strata of earth. And finally, by looking at our own DNA, they can learn so much about the evolution of our own ancestors.[45,46]

STORY

During the time of the "theoretical ancestor," all of Africa was mostly covered in tropical rainforests. However, due to tectonic shifts about six million years ago, mountains called the Great Rift emerged through the middle of Africa that separating east from West Africa while also changing the climate. West Africa stayed mostly the same as before, covered in rainforest, and most monkeys and apes still live there now. East Africa, however, changed considerably, becoming simultaneously hotter and drier as well as more unstable.[47]

Because of this, East Africa overall declined in trees and our theoretical ancestor gradually lost much of their ancestral foods—that is, fruits and leaves—and thus were forced to adapt by evolving into another species, called *Australopithecus*. Scientists have identified seven different species of *Australopithecus*—all with variations in their diet as evidenced by differences in their teeth and jaws, with some deemed more gracile and others more robust. Instead of focusing on one species, however, I will outline the general traits that applied to most of them for the sake of brevity.

ADAPTATIONS

Due to these changes to their climate and food, our ancestors attempted to adapt through inherent flexibility. When fruit was available, they continued to feed in the trees, but when fruit was scarce, they knuckle-walked over the ground to cover longer distances to acquire their fallback foods like tubers, grains and berries, as well as insects, shellfish, reptiles, rodents and other smaller animals. Since they were more exposed to predation by the large carnivores of the time, they walked short distances upright to scope for predators while staying more closely packed into groups and sleeping in caves and other enclosed areas at night.

They likely, too, adapted through epigenetics by becoming more hyper-alert: some studies show that animals born into stressful environments carry more stress hormones.[48] And they adapted through cultural evolution: like the Savannah Chimps, they probably invented digging sticks to unearth tubers and rocks to crack nuts, and then taught their offspring the same.[49]

But even then, they were not adapted enough to find enough food and save themselves from predators, so they started the long and arduous process of natural selection. As noted, these ancestors already knuckle-walked on the ground—and even walked upright for short distances. But some of them generated genes to allow them to do this even better, which in turn allowed them to cover longer distances and acquire more refined foods—and thus reproduce more and spread their genes into the future. And after many iterations of this, our ancestors developed bipedalism.

While climbing and knucklewalking, our ancestors were pushing against gravity with their muscles. However, with bipedalism, they were pushing mostly with their bones, using their muscles only for balance, so their locomotion was more efficient. They could also better spot predators while higher off the ground, and better dissipate heat because less of their body was exposed to direct sunlight. In the process, too, they became slightly taller than the common ancestor but considerably lighter, presumably because they did not need as much muscle for their locomotion. On average, Australopithecus stood about four to five feet tall and weighed between 65 to 100 pounds,[50] while chimpanzees are on average around four feet tall and weigh between 70 to 130 pounds.[51] All of this combined allowed them to travel longer distances with less energy, risk, and overheating, to collect their more refined and scattered foods from the ground, trees and even lakes and streams.[52]

They also evolved in other ways through natural selection. Because these ground foods were tougher and more brittle, their teeth and jaws became bigger and stronger as compared to their predecessor. They also reduced the size of their gut and perhaps most of all their colon, while their brain remained about the same size.[53,54]

IMAGE 11.1- AUSTRALOPITHECUS

CATABOLISM

Australopithecus continued to consume much of their carbs from fruit when available—but also consumed increasing amounts of starches from the ground, including tubers, roots and possibly grains.[55] Since starches contain more glucose, they consumed more glucose relative to fructose, at about 70% glucose to 30% fructose; and since fructose is converted into glucose, they thus performed fewer conversions. Also, because they consumed fewer plants and more animals for protein, they reduced their amount of fiber, allowing their colons to shrink, and thus reducing colonic fats to less than 10%—even as they continued to consume large amounts of fiber. They also consumed more dietary fats from animals as mentioned, which were also used for catabolism.

ANABOLISM

Given that they lost much of their principal sources of protein, tender leaves, *Australopithecus* consumed more animals,[56] and whole animals especially, than their predecessor, including insects, rodents, reptiles, shellfish and even larger animals. So they thus received protein that was lower in toxins and fiber, easier to digest, while containing more balanced ratios of amino acids, including both muscle and collagen protein. Through preferring fattier tissues, like the adipose, tongue, brain, eyes and marrow, they also received greater amounts of animal fats, with balanced ratios of all fatty acids, and an abundance of the longer chain polyunsaturates, AA, EPA and DHA. They attained all fatty acids needed for the anabolism of their phospholipids, especially ones in their brain and other nerves.

METABOLISM

To understand more about the nutrition of *Australopithecus*, let's consider that animals contain four types of metabolism: basal, digestive, exercise, and Total Energy Expenditure (TEE). Basal metabolism happens throughout our body when we are resting or sleeping. During this time, our hearts are beating, our lungs breathing, our livers and kidneys filtering and converting, and even our muscles are metabolizing at reduced rates. Digestive metabolism happens in various tissues when we are digesting our food, which is over and above basal metabolism. Exercise metabolism results from movement—that is, when our muscles move and thus require more metabolism. All these metabolisms combined are TEE, or the total amount of metabolism over one day.

Our common ancestor and *Australopithecus* had similar basal metabolisms because they were similar in mass, because all animals of similar size tend to have similar basal metabolisms. Since smaller animals lose more heat due to increased surface area relative to over-all mass, they have higher metabolisms to maintain body temperature; and the reverse for larger animals — thus in the end, animals of the same size trend towards the same basal metabolism.[57] But Australopithecus was probably slightly higher in basal due to their smaller muscles and guts, making them thinner overall while dissipating more heat. Due to their more refined diets and smaller guts, *Australopithecus* had less digestive metabolism. They probably had about the same amount of exercise metabolism because while their locomotion was more efficient, they covered more territory.

Generally, too, animals of the same size tend to utilize similar amount of protein per lean mass because their cells turn over protein at similar levels.[58] But if one animal exercises more than the other, they tend to turn over protein faster. Due to their smaller mass, *Australopithecus* likely needed slightly less protein and possibly consumed less protein, because they consumed higher-quality proteins. Since their brains were only slightly larger than their predecessor, they probably just needed slightly more glucose.

Taking all this into consideration, I estimate that *Australopithecus* probably consumed about the same number of calories per day as the common ancestor: about 1900. They probably consumed about 20% of their calories from protein (48 g), but also probably ate higher-quality proteins, which then allowed them to eat less overall protein; 50% of their diet from carbs (235 g) and 25% from dietary fat (52 g), and less than 10% from colonic fats.

FOODWAYS EQUATION

As compared to the theoretical ancestor, Australopithecus, on the calories side of the equation, consumed about the same amount of calories, maybe slightly more. But they consumed more refined diets, thus improving the quality and balance of their macronutrients — but only slightly. By eating more animals, they improved the quality of their proteins and fats — and avoided more secondary compounds from plants. By eating more starch, they improved the quality of their carbs by consuming more glucose relative to fructose. And they received more calories from carbs and dietary fats relative to colonic fats—which delivered more ATP.

On the metabolism side of the equation, they kept their metabolism, their TEE, about the same because their bodies and brains were about the same size. Sensing and ingestion remained about the same. Locomotion and capture, possibly, stayed about the same. However, they shrank the metabolism of their digestion due to smaller guts and performed fewer conversions in their liver. However they did slightly increase the size and thus the metabolism of their brain—which, in turn, helped them find their more refined foods in their increasingly challenging environments.

FOODWAYS EQUATION OF *AUSTALAPITHECUS* COMPARED TO THE COMMON ANCESTOR

CALORIES: about the same: 1900 or slight increase.

=

MACRONUTRIENTS: increased refined foods mostly by eating more animals.

=

Protein: same as percentage of calories: 20%.
Completes: increased from more animal foods.
Collagen: increased from more animal foods.

+

Carbs: decreased as a percentage of calories: 55%.
Fruits: decreased: 35%.
Starches: increased: 15%. Consumed more glucose relative to fructose.

+

Fats: increased: 25%.
Dietary: increased: 20%.

Colonic: decreased: 5%.

=

METABOLISM: about the same

Sensing: same

+

Locomotion: increased in efficiency but covered more ground.

+

Capture: (hunting and gathering): increased in effeciency but covered more territory.

+

Ingestion: increased due to rougher plant foods from the savannah such as tubers and seeds. This contradicts the usual trend towards smaller teeth and jaws in ominivorous primates.

+

Digestion: decreased, paradoxically, due probably to more animal foods.

+

Liver: (conversions and detoxification): decreased to more balanced macronutrients.

+

Brain: (cognition and sociality): increased slightly, maybe

CHAPTER 12

HOMO ERECTUS

STORY

About two million years ago during the Ice Age, otherwise known as the Pleistocene, the climate in East Africa continued the overall trends towards hotter and dryer while also fluctuating back to cooler and wetter, creating dual challenges for adaptation. With trees even more in decline along with fruits, *Australopithecus* could not walk long enough distances to acquire enough food, especially given their susceptibility to predation. With their smaller brains, they could not use cultural evolution to adapt to the constant changes in their environment. So they were forced to evolve or perish. Likely, one species of *Australopithecus*, either *afarensis* or *africanus*, evolved into another hominoid called *Homo habilis* and then another *Homo ergaster*, who then evolved into *Homo erectus*. *Homo erectus* appeared about two million years ago around lakes and rivers in East Africa. This new species likely drove *Australopithecus* into extinction by outcompeting them for territory and food—and then eventually migrated across other parts of Africa into the Levant as well as Eurasia where they indirectly evolved into Denisovans and Neanderthals.

They likely continued the same or similar social structures as the common ancestor and *Australopithecus*. However, some studies show that they had less sexual dimorphism than their predecessors, suggesting even greater cooperation and equality between males and females. They probably camped in caves, and other protected places or crude structures likely around fires while hunting and gathering in smaller groups during the day.

Homo erectus adapted to the changes in their environment through natural selection in four ways. First of all, they adapted their loco-motion, downregulating their climbing and upregulating their bipedalism even more, to allow them to cover more distance even as they could still climb trees reasonably well. Secondly, they evolved smaller teeth and jaws, as well as much smaller guts, because of their more refined and processed or predigested diet. Thirdly, they evolved larger bodies possibly to help them outcompete *Australo-pithecus*, protect from predators, generate more heat, and use longer legs to cover greater distances. While *Australopithecus* was about four feet tall and eighty pounds on average, *Homo erectus* was 5.5 feet tall and 120 pounds on average.[59] Fourthly, and most of all, they evolved much larger brains, perhaps to adapt to the increasing fluctuations and challenges in their environment. While *Australopithecus's* brains were about 400 cubic centimeters, *Homo erectus* brains were about 1000 cc, so they more than doubled the size of their brain, coming close to our own size of 1300 cc.[60] They looked less like us, with long arms and legs, slender bellies and smaller heads.

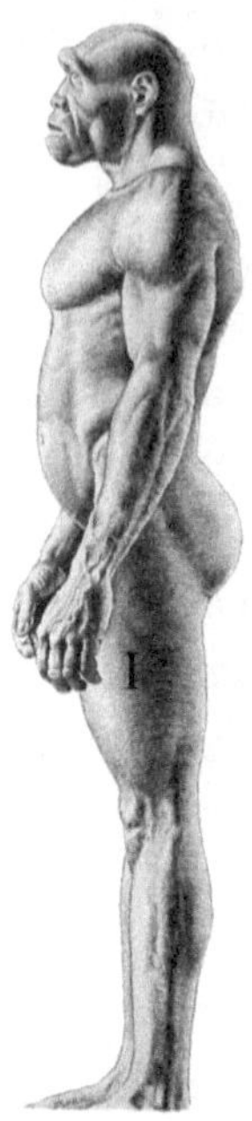

IMAGE 12.1- HOME ERECTUS

ADAPTATIONS

With their larger brains, *Homo erectus* radically changed the way they adapted to their environment—that is, through cultural evolution. When their foodways were disrupted, they did not need thousands of years to adapt through natural selection by changing their morphology: they could adapt more instantaneously, through using their intelligence and creativity and then hand those adaptations down to their offspring through educating them. Instead of evolving claws and fangs for hunting they merely made them from sticks and stones. While their predecessors used tools in crude and simple ways—smashing rocks and digging sticks for example—*Homo erectus* used them in more advanced ways. They used rocks to make weapons, mostly hand axes and the like, and used them for hunting and butchering animals and for smashing and grinding plants. They used spears, too, for hunting and protecting themselves from rivals and predators. They likely used crude scythes to harvest grains, and stone hoes and digging sticks to unearth tubers.

With their greater cultural evolution, they also used strategies to acquire their food, especially in hunting. They used persistence hunting to acquire larger animals like ruminants.[61] With their thin, upright and less hairy bodies they could cover more ground and dissipate heat better than most large mammals—which in turn allowed them to chase down those animals over several hours and kill them with their tools. They likely used many other strategies for hunting, including ambushing and stalking while also working in groups to surround herds of animals and drive them over cliffs or into ravines and lakes for harvesting.

Through hunting larger animals and then preserving that food for later consumption, *Homo erectus* greatly increased their supply of food. Obviously larger animals can provide about 100 times as much food as smaller ones; however, they do not necessarily require that much more metabolism to hunt. So, in hunting larger animals, *Homo erectus* could acquire way more calories with way less metabolism in locomotion and acquisition. At the same time, they could then preserve the meat and fat from these animals through drying, smoking and rendering. So, for the same amount of metabolism, *Homo erectus* radically increased their amount of food, altering their foodways equation.

With their greater intelligence, *Homo erectus* likely used fire for protection, light and cooking—which then separated them from nearly all other animals.[62] While nearly all other animals eat hand to mouth, *Homo erectus* acquired their food, delayed eating that food, carried their food back to their village, and then predigested that food through smashing, cutting, grinding and cooking. Most interestingly, they then shared that food with others in their group, showing the enormous steps in intelligence and, most of all, sociality and cooperation. Also, through cooking, *Homo erectus* consumed more plant foods because cooking deactivates many antinutrients and toxins in plants. Additionally, they could probably preserve and store more of their plant foods. So overall they needed less locomotion and metabolism to acquire their plant foods.

METABOLISM

As compared to *Australopithecus*, the basal metabolism of Homo erectus was considerably higher due to their larger bodies and brains—even when calculating for thermogenesis as explained in prior sections.[63] They decreased digestive metabolism substantially due to their refined diets and smaller teeth and guts. They probably kept exercise metabolism about the same: while more efficient in locomotion, they nonetheless traveled greater distances. But overall they increased their TEE (or total metabolism) considerably.

As a percentage of overall calories, protein probably stayed about the same or possibly decreased slightly because, by eating more animal as compared to plant proteins, they increased the quality of their protein and thus could eat less protein overall. They probably consumed slightly fewer carbs but more of them from starch and thus glucose. And they substantially increased their consumption of dietary fat as compared to colonic fat mostly by eating more animal fat. So from their predecessor they increased their calories from 1900 to 2375 overall, with 20% of those calories from protein or 118 g; 45% from carbs or 267 g; and 30% from fat or 80 g; and less than 5% from colonic fats.

HORIZONTAL AND VERTICAL

Before moving into my conclusions on foodways, I need to introduce another concept into my theory—that is, the concept of expensive tissues. Some tissues are inexpensive, meaning they do not require many calories per hour for their metabolism. Some are expensive, meaning they require more calories per hour. When considering organs in our metabolism, we see the following results per day for males: our brains use about 200 calories; our liver about 240 calories;

our kidneys about 150 calories; our guts about 200 calories, depending on our diet; our heart about 180 calories. Our muscles vary enormously, however, with our activity or exercise metabolism: ranging from about 80 calories per hour at rest and then from 150 to 1000 calories per hour with different activities. From now on, I refer to tissues that require more metabolism as "expensive."[64]

In our evolution, our ancestors increased the size of one of the most expensive tissues, our brain; while 2% of our bodyweight, our brain uses 20% of our calories and nearly all of that as glucose. But we also decreased the size of some of our most expensive tissues, including our guts and livers and perhaps kidneys as well.

FOODWAYS EQUATION

On one side of the foodways equation, Homo Erectus increased the amount and quality of their calories; on the other side, they also increased their brain metabolism but decreased their gut and liver metabolism; they also probably decreased their locomotion and capture metabolism as well. But in the end, they balanced the equation as usual.

From here on out, on the foodways equation, I will note some tissues as "expensive."

FOODWAYS EQUATION OF *HOMO ERECTUS* COMPARED TO *AUSTALOPITHECUS*

CALORIES: increased substantially: 2400.

=

MACRONUTRIENTS: increased refined foods from eating more animals, especially animal fat, and further predigestion and cooking of all foods.

=

Protein: same or lower as percentage of calories: 20%.
Completes: increased from more animal foods.
Collagen: increased from more animal foods.

+

Carbs: decreased as the percentage of calories: 50%.
Fruits: decreased: 25%.
Starches: increased: 25%.

+

Fats: increased: 30%.
Dietary: increased: 25%.
Colonic: decreased: 5%.

=

METABOLISM: increased substantially.
Sensing: same.

+

Locomotion: increased efficiency but maybe covered more ground.

+

Capture: (hunting and gathering): decreased due to tools and strategies.

+

Ingestion: decreased due to more refined and predigested foods.

+

Digestion: decreased for the same reason, expensive.

+

Liver: (conversions, storage and detoxification): decreased, expensive.

+

Brain: cognition and sociality: increased dramatically, expensive, which in turn allowed them t

Ingestion: decreased due to more refined and predigested foods.

+

Digestion: decreased for the same reason, expensive.

+

Liver: (conversions, storage and detoxification): decreased, expensive.

+

Brain: cognition and sociality: increased dramatically, which in turn allowed them to acquire their more refined diet, expensive.

CHAPTER 13

HOMO SAPIENS FORAGERS

STORY

Somewhere around 400 to 300 thousand years ago, Africa was possibly experiencing some of the most extreme weather of the Pleistocene, with extreme heat and aridity as well as fluctuations, causing much of Africa, and especially East Africa, to turn into desert. During this bout, *Homo erectus* encountered too many obstacles to their survival and therefore started their evolution towards *Homo sapiens* in East Africa, probably around lakes and rivers. For most of the next 200 thousand years or so, *Homo sapiens* were somewhat marooned in this part of Africa, surrounded by deserts, but during times of moisture, they migrated into northern and likely other parts of Africa.

Around 60,000 years ago, they migrated all the way into South Africa, living in caves along the coast. Finally, around 50,000 years ago, they migrated further through Africa and then across the Red Sea into the Levant and then Eurasia, where they encountered the descendants of *Homo erectus* who migrated there earlier, including Denisovans and Neanderthals. During the Glacial Maximums, they stayed in southern Eurasia, but during the retreats they ventured further north. At the start of the Holocene about 10,000 years ago or sooner, they ventured all the way into Siberia, probably following herds of reindeer, and finally across Beringia into Alaska and then into the Americas.

ADAPTATIONS

Homo sapiens remained mostly the same as *Homo erectus* except for three adaptations through natural selection that, however, are extensions of the same trends that differentiated *Homo erectus* from *Australopithecus*. First, they evolved even smaller jaws, teeth and guts because they were consuming more refined and predigested foods. They were about the same height as *Homo erectus* but with thinner skeletons and more robust bodies, so while *Homo erectus* was about 120 pounds on average, humans were about 150 pounds. *Homo sapiens* probably became larger to improve their hunting of large game, to preserve heat, and fight against each other. And most importantly, they evolved larger brains, about 30 to 40% larger than *Homo erectus*, with most of that probably devoted to cognition and sociality[65]—all probably for the same reason as before: to deal with changing and new environments as well as the times of intense dryness and scarce food.

Additionally, humans also evolved differences from each other to adapt to their various environments, primarily in two different ways. First, they varied the color of their skin, mostly to regulate vitamin D and to protect from sun damage. So when closer to the equator, they evolved darker skin while still allowing for the synthesis of vitamin D. When further from the equator, they usually evolved lighter skin to collect more sun—even though some, in the far north, maintained darker skin because their food contained vitamin D, mostly from animal fat and liver.[66]

Second, they varied both the overall mass as well as the length and thickness of their body[67] to help with thermoregulation, locomotion, and nutrition. For example, at one of the extremes, humans in jungles evolved less mass overall to account for less food, especially less fat, and to better dissipate heat; and they became shorter for that same reason, but also to enhance locomotion through dense foliage. Meanwhile arctic people are squatter and even shorter to conserve heat and to store more calories through long winters.

But once they evolved their larger brains, *Homo sapiens*, just like *Homo erectus*, then adapted mostly through cultural evolution[68]—not just to one ecosystem but many, as they migrated back into jungles, then into coastlines, deserts, woodlands, grasslands and even tundra. They invented not only better tools for hunting, but also many types of tools for hunting different animals in different environments, including axes, knives, spears, harpoons, slingshots, atlatls, and later other projectiles like bows and arrows. They also invented more sophisticated tools for gathering plants, including hoes, scythes, sacs and the like.

Homo sapiens also invented better strategies, involving cooperation and communication, for hunting animals, including surrounding whole herds of animals and then driving them over cliffs, into ravines or lakes and then slaughtering them. They also enhanced their innovations to predigest food through multiple methods including grinding, fermenting, and various forms of cooking; and enhanced their ability to store food through drying, smoking, curing, freezing, rendering, as well as storing in sealed and airtight animal skins.[69] In other words, they continued to push the same cultural evolution of *Homo erectus* to the next level—which in turn allowed them to adapt to just about every ecosystem on earth, whereas nearly all other animals just adapt to one or possibly several ecosystems.

UNIVERSAL HUMAN DIET

While evolving from the common ancestor, into *Australopithecus*, *Homo erectus* (with several intermediary species along the way), and finally *Homo sapiens*, our ancestors followed most of the same trends on the calories side of the foodways equation—the same trends that also extend through omnivorous monkeys and apes. We increased our overall consumption of calories and improved the quality of our macronutrients by consuming more animal proteins and fats and consuming more plants higher in glucose relative to fructose.

We also increased our consumption of dietary fats and lowered our colonic fats.

Additionally, all our ancestors increased their basal metabolism because they became larger even as they decreased it per mass for thermoregulation. They decreased their digestive metabolism because they evolved towards even more refined and predigested foods. They likely decreased their exercise metabolism overall because they became more efficient with bipedalism. And they decreased their locomotion and capture because they could consume greater amounts of food at one time, especially large, fatty mammals—and then store those nutrients for later consumption. And they radically increased their brain metabolism by three times. Given all of this, we increased our TEE, total energy expenditure, overall.[70]

This long evolution, from the common ancestor to foraging humans, trended us towards what I call the Universal Human Diet—the diet that most of us are most evolved to consume, on average, at least while living as foragers.

CALORIES: 2800 per day for males, extremely active compared to moderns.

PROTEIN: 20% of calories (140 g) mostly from animals but also plants.
Completes: 80% of total protein.
Collagen: 20%.

CARBS 35% of calories (245 g) from whole tubers, fruits and grains.
Glucose: 70% of total carbs.
Fructose/Galactose: 30% of total carbs.
Fiber: 40 g, about half soluble or fermentable fiber

FATS: 45% of total calories, mostly from animals but also from plants.
Saturates: 35% of total fat.
Monounsaturates: 45%.
Polyunsaturates: 20%.

Two to ten parts omega six to one part omega three as shorter and longer chains.

Lots of preformed vitamin A and B12 from liver and phytonutrients from plants.

I call this the Universal Human Diet for several reasons. First of all, our ancestors clearly trended in this direction. Second, most human cultures, past and present, trend towards this diet as well whenever all the nutrients are available to them—that is, foraging, agricultural, and modern cultures. I am not aware of any culture that eschewed one class of macronutrient when they were available to gorge themselves on one or two of the others. Third, this diet accords with our nutritional physiology, giving us all the nutrients we need in their optimum ratios while minimizing conversions—as was addressed in a prior section of this book. However, this diet does assume the same level of exercise metabolism of foragers and agriculturalists, with foragers in particular locomoting about seven miles per on average while squatting for rest which kept their muscles engaged and metabolizing. So as we shall see, this diet does not necessarily apply to humans with their reduced exercise metabolism.

VARIATIONS IN THE HUMAN DIET

When inhabiting various ecosystems, our ancestors could not always find enough foods or the right kind of foods to conform to the universal diet, so they deviated from that diet. In some cases, they could not find enough food, so they ate fewer calories overall and lived with permanent malnutrition or shrank in stature.[71] In almost all cases humans could acquire lots of protein in any ecosystem because they were the apex hunters, but without enough fat and carbs, they could not eat all of that protein because humans can tolerate only so much protein before experiencing detriments and even death.

Warmer ecosystems

In many warmer and tropical ecosystems, foragers could acquire plenty of carbs as these environments contain fruits, tubers, grains, berries, and honey. But they are usually low in plant fats and animal fats because animals do not need to store as much fat for insulation and food during longer winters. Foragers thus generally consume lean animals, some fatty plants like the balboa and mongongo, but larger amounts of carbs from tubers, fruits, honey and grains. They also consumed enormous amounts of fiber, perhaps as high as 100 g or more per day, as much as Chimps. Generally the following cultures ate diets like this, including the Hadza, Kung, the Aboriginals, Tsimane, Yanomami, and likely the Natufians and Kebarans during the Pleistocene—all who tended towards smaller and thinner bodies. Hadza adults, who live in tropical and arid Africa, generally consume about 2500 calories per day: 24% protein (although this sounds high to me), 65% carbs, 11% from dietary fat and 10% from colonic fat. The Kung consume about 15% protein, 60% carbs, 20% dietary fat and 5% colonic fat.

IMAGE 13.1- THE HADZA

Colder ecosystems

In colder environments, foragers consumed more fat than carbs for their calories. While colder environments produce few carbs, they produce enormous herds of fatty animals, both from the land and water, that store fat for insulation and food during the longer winters. So, these foragers consumed enormous amounts of protein and fats and, in some cases, they also consumed at least enough carbs for their needs. In other cases, they only consumed minimal amounts of carbs and thus performed permanent glucose and ketogenesis in their bodies. When migrating from Africa into Eurasia, our ancestor, Cro- Magnon Man, lived in grasslands and tundra, especially during Glacial Maximums, while consuming fatty animals like reindeer, mammoths, aurochs, and buffalo who fattened themselves on grasses, lichens and shrubs.[72] During Glacial Retreats, they probably consumed enough carbs; other times not.[73]

The Plains Indians of North America hunted bison in the fall when they were fattest and made pemmican rich in tallow. Some have estimated their diet was about 21% protein, 58% fat, and 22% carbohydrates.[74] The Inuit live mostly along the coast in the arctic and consume marine animals like seals, whales, fish and small birds, and only trace amounts of plant foods in the summer. They consume 30% protein, 50% fat and less than 20% carbs;[75] interestingly, they receive these carbs from select animals, like seals, that carry lots of glycogen in their muscles and liver, which in turn allows them to hold their breath and dive deep into the ocean.[76]

IMAGE 13.2- THE INUIT

RATIOS OF MACRONUTRIENTS

Across all ecosystems foragers generally consumed about the same ratio of amino acids in their diet, assuming they consumed mostly animals for protein. Animals contain similar ratios of amino acids in their muscles and organs, and similar ratios in their collagenous tissues. So, whether eating gazelles or reindeer, or fish and rodents, humans were consuming the same approximate ratios of amino acids. However, some animals, specifically smaller and thinner ones with more joints and long appendages, probably contain more collagenous proteins than found in larger and thicker animals. For this reason, it seems that many foragers, who specialized on larger animals, also supplemented with smaller animals, including small birds, tiny fish, rodents, and insects. But in general, it seems that most foragers ate their amino acids in similar ratios to each other.

Across various ecosystems humans also consumed similar ratios of sugars—that is, ratios of glucose to fructose. In almost all cases foragers consumed more starch relative to fruit, berries, and honey. But in warmer environments, they tended to consume more fruit and thus less glucose relative to fructose—and perhaps the reverse for colder environments. But likely in all cases humans generally consume somewhere around 60 to 90% of their carbs from glucose and the remainder from fructose.

VARIATIONS IN FATTY ACID RATIOS

Across ecosystems, foragers, however, consumed different ratios of fatty acids—because they ate different types of fatty plants and animals, which in turn contain varying ratios of fatty acids. Plant fats are usually low in saturates, high in monounsaturates and polyunsaturates, usually from omega-6. But animal fats are high in saturates with about equal parts monounsaturates and low amounts of polyunsaturates. However, some animal fats, especially from smaller animals like birds, rodents, and fish, contain lower amounts of saturates compared to monounsaturates and higher amounts of polyunsaturates. Foragers across various ecosystems ate various types of fatty animals, sometimes mixing fatty plants with fatty animals, sometimes not—and as such they ranged in their ratios of fatty acids.

Warmer environments

As noted before, foragers who lived in warmer environments tended to eat less fat and more plant foods relative to animal fats. So they ate more plant fats and animal fats from smaller animals. So overall they tended to consume less saturated fat, more monounsaturated fat and more polyunsaturated fat as omega-6, while always consuming some amount of omega-3. Since they ate more fibrous plants, they produced more colonic fats. This included many of the cultures in warmer environments already mentioned, such as the Hadza, who ate their ratios of fats found in the Universal Diet.

Colder environments

In colder environments foragers ate more fats relative to carbs and specialized on large fatty animals like hippos, reindeer, woolly mammoths, seals, and whales, while at the same time supplementing with smaller animals. In most cases, it seems they did not consume many, if any, plant fats, but probably did whenever they were available. This includes the cultures already mentioned, including Cro Magnon, the Plains Indians, the Inuit and many others. As compared to the Universal Diet, they tended to consume more saturates, fewer monounsaturates and considerably fewer polyunsaturates with closer ranges of Omega-6 to -3.

The Inuit, in particular, consumed mostly marine animals that are equal parts saturates, monounsaturates and polyunsaturates, with nearly all those polyunsaturates as omega-3. But at times the Inuit also consumed enormous amounts of small birds, rodents, and reindeer from the interior—so overall they probably consumed about 30% saturated, 40% monounsaturated and 30% polyunsaturated at maybe equal parts omega-6 to -3. They are possibly specially adapted to this ratio of omega-6 to -3 through natural selection.

While humans varied considerably in their ratios of fatty acids, those variations, however, were not all that pronounced and usually stayed within these ranges:
Saturated: 25 to 45%.
Monounsaturated: 40 to 45%, possibly as high as 60%.
Polyunsaturated: 10 to 30% with extremes of 5 to 40%.
Omega 6 to 3 ratio: 2 through 10 to 1 (although the Inuit consume more omega three than six).

The polyunsaturates were somewhere around two to ten parts omega-6 to one part omega-3, with one extreme of the Inuit who consume more omega-3 to -6. If some foragers consumed lower percentages of the essential fatty acids, the polyunsaturates, relative to other fats, they nonetheless attained enough of them through eating so much fat—while also eating animal tissues high in the omegas, most of all the brain, eye and spine. Curiously, these norms, if you remember, also work with our nutritional physiology and correlate to the way our bodies store fatty acids in our phospholipids and our adipose.

FOODWAYS EQUATION

On one side of the equation, *Homo sapiens* increased their calories through using cultural evolution to acquire more food—and then store that food for the long term. They also continued the trend towards higher quality proteins and fats—as well as more starch relative to fruit. But ultimately they increased their calories mostly from the consumption of animal and plant fats which, as we know, provide enormous amounts of steady ATP. On the other side of the equation, they also increased their metabolism overall due to larger bodies and brains. But at the same time, they also increased and decreased various parts of their foodways as explained below. Most importantly, through increasing their brain metabolism, they evolved enough cultural evolution to acquire enormous amounts of food in just about every ecosystem on earth and then process, store and share that food.

FOODWAYS EQUATION OF *HOMO SAPIENS* COMPARED TO *HOMO ERECTUS*

CALORIES: increased: 2800 per day.

=

MACRONUTRIENTS: increased refined foods through more animals and predigestion.
Protein: same as percentage of calories: 20%.
Completes: 15% mostly from animals.

Collagen: 5% from animals.

+

Carbs: decreased: 35%.
Fruits: 10%.
Starches: 25%.

+

Fats: increased substantially: 45%.
Colonic: decreased: less than 5%.
Dietary: increased: 40%:
30% saturated.
45% monounsaturated
25% polyunsaturated: six parts of omega-6 to one part omega-3

=

METABOLISM
Sensing: same.

+

Locomotion: decreased due to more efficient movement, hunting, gathering and storage, but also increased for more territory

+

Capture: (hunting and gathering): decreased for the same reason.

+

Ingestion: decreased due to more refined diet.

+

Digestion: decreased, *expensive*.

+

Liver: (conversions, storage and detoxification): decreased, *expensive*.

+

Brain: (cognition and sociality): increased dramatically, *expensive*.

One interesting question is: what macronutrient drove our evolution towards humans? Or put another way, was there some macronutrient they acquired that allowed them to continue their evolution all the way into humans? As noted, our ancestors increased their metabolism so they needed more macronutrients — but which ones?

Obviously they needed more protein for bigger bodies, but they also needed less per unit of mass. At the same time, *Homo erectus* was already an apex hunter, able to kill many large animals at once. So they could already acquire enough protein for their needs—so protein was likely not the bottleneck or rate limiter.

Obviously, too, our ancestors evolved to catabolize more glucose for their larger brains—so we might assume more carbs drove our evolution. Our theoretical ancestor needed about 50 g per day for their brains while human need about 150, so we evolved to need about 100 g more. But our theoretical ancestor already consumed about 250 g of carbs per day from fruit. Furthermore, we did not evolve to consume more carbs but less and, accordingly, our environments contained fewer carbs. So, we evolved, paradoxically, to catabolize more carbs but eat less of them, especially when adjusting for our increased mass.

So what about fat?

Our ancestors likely acquired their additional calories from fats—from plant and mostly animal fats: from specifically hunting large, fatty animals and then rendering and storing their fat for later use—thus providing steady sources of enormous amounts of ATP over the long haul. Paradoxically, our brain does not catabolize fats; the long chains usually do not even pass through our blood-brain barrier. However, these long chains provided enormous amounts of ATP to all other tissues, the ones that prefer these fats—and that spared more glucose for our brain and provided calories for our larger bodies.

However, Africa did not contain many sources of plant or animal fats.

As noted before, even during Glacial Maximums, Africa was still mostly tropical so that animals did not need to store much fat for insulation or food. East Africa does contain some fatty seeds, such as nuts, but not many. So this raises the question: where did *Homo sapiens* find enough fat in Africa to sustain themselves?.

Homo Sapiens evidently settled into only the select and limited ecosystems in Africa that contained fatty animals. In East Africa some of the earliest humans, called the Omo and Herto, settled around lakes and rivers, full of hippos loaded with fat: even one kill could feed them for months if the meat was dried and the fat rendered. Given that hippos are so ferocious, they possibly evolved from *Homo erectus* into *Homo sapiens* while hunting these animals.[77,78] In South Africa, the Klasies and others settled along the coastlines loaded with Cape Fur seals and the occasional beached whales and probably other sources of marine fat.[79]

When migrating northward out of Africa, *Homo sapiens*, then called Cro Magnon Man, settled into cold and even arctic ecosystems in Eurasia full of fatty mammals, including reindeer, aurochs, wooly mammoths and many others. Others settled in small populations in the Levant with plant fats from olives and nuts while also hunting fatty sheep in the mountains. So it appears, based on all evidence, that dietary fat drove human evolution towards greater mass and intelligence. Through this source of steady calories, we then attained more calories than needed for our metabolism, allowing for one of our most unique traits: leisure.

CHAPTER 14

HOMO SAPIENS: AGRICULTURALISTS

STORY

Starting somewhere around forty thousand years ago during the Ice Age, humans migrated from Africa into the Levant and then into Eurasia. But small amounts of them stayed in the Levant and foraged for foods that were later domesticated in that region, including wheat, barley, peas and lentils, as well as goats and sheep. During glacial maximums they, and specifically the Kebarans, were nomadic, moving from one place to another with the seasons.[80]

But during one glacial retreat and global warming, another group, the Natufians, became sedentary, staying in the same villages year-round while practicing rudimentary forms of farming like sowing grains on fertile lands.[81] However, during the Younger Dryas, when temperatures turned colder again, they became nomads. But then again, when the Pleistocene shifted about 10,000 years ago into the Holocene, into more permanent and stable warming that defines our modern era, they converted over into full agriculture, starting with just farming and then adding herding. Eventually, many then settled behind defensive walls in the village of Jericho.

Interestingly, agriculture also developed several thousand years later in other parts of earth, independently[82]—that is, without those cultures having any contact with the Levant. These areas included China and the Americas about 7,000 years ago, as well as New Guinea and parts of Africa somewhat later. Additionally, once initiated in any of these regions, agriculture then spread aggressively outward.

Scientists continue to debate the reasons why humans became agriculturalists. But to me the most likely scenario is this: During colder and drier eras in the Levant, humans foraged for grains and legumes as mentioned, but nonetheless relied heavily on fatty animals, especially the sheep that inhabited the mountains. Even if farming were possible during that time, they would probably abstain because hunting fatty animals was more efficient than farming carbohydrates—and more nutritious and tastier. Furthermore, humans tend to prefer foraging to agriculture when given the chance and only revert to agriculture when necessary. But even considering all of this, these foragers generally could not find that much food in this area and thus they remained scattered and low in numbers for tens of thousands of years, usually located within specific ecosystems around lakes or rivers.

However, during the warmer periods, the grains and legumes proliferated more while fatty animals went into decline or otherwise became leaner. So these foragers acquired more of their calories from plant carbs relative to animal fats—and thus became dependent upon them for their survival. Accordingly, they learned to make them more abundant—thus giving rise to agriculture. Because of their pre-existing knowledge of these plants, they soon learned to cultivate them in various stages. First, they just sowed grains along fertile soil close to their dwellings and harvested the crops later. They then started to till, fertilize, and weed the soil—and eventually, carefully select certain seeds for reproduction to enhance the qualities of that particular food. Overall they turned to agriculture because they could not acquire enough fatty animals anymore.

Once generating enough excess grains and their grasses, they then lured animals closer to their dwelling, caged or yoked them, and eventually domesticated them, too, through selecting the more docile and probably fattier animals. After some time, animals themselves became domesticated and tame, allowing for the herding of them in the wild without fences.

Overall, this region, too, unlike all other parts of the world, contained many of the plants most suitable for domestication, including grains, and specifically wheat and barley, as well as legumes, and specifically peas and lentils. Within six months of planting these seeds, humans could harvest them for consumption, whereas fruits, nuts and many other plants require years to yield food. Additionally, these plants are more abundant in macronutrients than just about any other foods on earth, containing lots of carbs, decent amounts of protein and the essential fatty acids, the polyunsaturates as both omega-6s and -3s. Additionally, when grains and legumes or nuts are combined, they create complete proteins—which prevents the malnutrition that would otherwise develop if humans specialized only on one or the other of these types of plants. This region, too, also contained the animals most suitable to domestication, namely goats and sheep, because they are small, docile and gregarious.[83] And even more, the Levant also contained varieties of ecosystems, close to each other, that allowed for the cultivation of different plants, thus creating backups in case one crop in one ecosystem failed. Once started, agriculture becomes self-replicating, as I will explain at the end of this section.

ADAPTATIONS AND PATHOLOGIES

In the above section, we can see the usual patterns that drove our evolution: the weather changes, which then changed the ecosystem, and then humans started to adapt through the usual methods. Through inherent flexibility, they turned to their fallback foods, grains and legumes and eventually exploited their preexisting knowledge of them. They probably adapted through epigenetics, turning certain genes on and off to switch from more animal to plant diets. But as usual for all humans, including foragers, they adapted through cultural evolution, which they then advanced to another level to domesticate plants and animals, using tools, strategies and even social structures. However, they also continued to evolve through natural selection to consume certain foods that became more abundant during agriculture, including dairy, salt, and alcohol, as well as to some extent wheat, legumes and possibly others.

However, they also adapted—or really just changed—often for the worse. More specifically, agriculturalists, as compared to foragers before them, changed their morphology, becoming shorter and weaker with thinner skulls and smaller brains.[84] They were also neotenous, meaning they maintained juvenile characteristics into adulthood, including smaller jaws, wider eyes, flatter faces and smoother skin.[85] Additionally, agriculturalists presumably became more docile, less aggressive, and less resistant to authority. Interestingly, when animals are domesticated, they follow these same patterns due mostly, in their case, to selective breeding. But in the case of humans, scientists cannot necessarily delineate what caused these changes. Epigenetics? Natural Selection? Even sexual selection? Or something altogether different: poor nutrition and lifestyle?

They possibly evolved through natural selection for smaller bodies and brains because that is more adaptive for agriculture. However, they were probably malnourished in childhood which tends to reduce stature in adulthood by as much as six inches, and this in turn might also cause the brain to shrink. In other words, they just could not reach their genetic potential. Epigenetics probably played some part in all of this as well.

In other cases, however, agriculturalists changed for the worse clearly due to their diet and lifestyle—which caused the prevalence of chronic and infectious diseases. Chronic diseases included tooth decay as well as bone and joint disorders, anemia, vitamin deficiencies, heart disease, and likely many other conditions that are rare or nonexistent in foragers. Agriculturalists were way more susceptible to infectious diseases due to their reduced health, crowded populations and proximity to domesticated animals who spread these diseases. These diseases included lesser ones like cold and flu as well as lethal plagues. They also suffered from various social diseases, including inequality, oppression, poverty, discrimination and, presumably, from various mental illnesses as well.

LIFESTYLE

Agriculturalists also changed their lifestyle. In the prior pages I surmised that our ancestors maintained about the same lifestyle from the common ancestor through foraging humans—but that all started to change, radically, during agriculture. Instead of living in the wilderness like foragers, agriculturalists increasingly destroyed the wilderness to expand their farms. Whereas once living in small groups, not exceeding one hundred, they now lived in larger villages, consisting of hundreds and even thousands of people, many of whom were strangers. Instead of hunting, men mostly herded and farmed and performed numerous other chores. Women continued to care for the children and their households as well as prepare meals, but instead of foraging for plants, they farmed in the fields. In general, both men and women, and perhaps men most of all, performed work that was way more monotonous, tedious, repetitive and lacking any adventure, challenge or novelty—while also working, possibly, about twice as much. Increasingly over time, agriculturalists became less egalitarian and cooperative and more hierarchical and oppressive, even dividing into classes with some ruling over others.

DIET

Early

Like foragers, agriculturalists did not all eat the same diet, but four different types of diets that I define as: early, privileged, unprivileged, and herders. In some cases they continued to live in small villages with more egalitarian and cooperative societies, while still surrounded by some amount of wilderness while eating both wild and domesticated foods. For protein they continued to consume mostly wild animals at first, as well as combinations of grains and legumes, and later domesticated sheep and goats. For carbs they domesticated only grains at first but soon added legumes, nuts and other seeds, such as sesame and flax, which generated complete proteins and even more essential fatty acids. They also continued to eat plants from the wild, including fruits, nuts and herbs. Due to their leaner animals, they consumed lower amounts of dietary fats and thus received more calories from carbs while also generating ample colonic fats. Even though seeds are abundant in antinutrients, toxins and storage proteins, they nonetheless processed them to reduce these compounds, including roasting, soaking, germinating, fermenting, filtering and cooking. Generally, these people seemed healthier than many agriculturalists and probably consumed about 20% protein, 50% carbs, mostly from starches and some wild fruits and berries, and maybe 30% or lower as fat and 5% of that as colonic fat. These people came before the introduction of the new foods listed below.

Grains, Legumes, and Nuts

Before describing the next three categories of diets, I first wanted to address the new foods, or foods that were eaten in higher quantities, that were introduced later in agriculture, including grains and legumes, fermented foods, alcohol, dairy, salt and eggs.

Although we tend to think that our ancestors did not start to consume grains and legumes until agriculture, they probably started as far back as the common ancestor based on several lines of evidence. Grains and legumes existed in Africa concurrent with some or all of our evolution, including millet, sorghum, and teff, as well as several legumes, including bambara and cowpea. Secondarily, they are the most nutritious plants on earth, even when wild. At the same time, they are covered with mechanical and chemical defenses—but our ancestors could overcome those defenses. And indeed we have concrete evidence that our ancestors were consuming wild sorghum in Africa back 100,000 years ago.[86]

As already noted, humans in the Levant were consuming wild wheat and barley, as well as peas and lentils, for tens of thousands of years before agriculture. However, during agriculture, we probably consumed way more of these foods than ever before, because we could produce an abundance of them. Some evidence suggests that we evolved through natural selection to these foods during this time, to help us better digest starch and better absorb iron, and alter our microflora to ferment the fiber in these foods.[87]

Fermentation

But our ancestors principally adapted to these seeds through cultural evolution—that is, they invented ways to to reduce the antinutrients and toxins in them while also predigesting them. For starters they cultivated their seeds to contain fewer secondary compounds by selecting and breeding ones that were less bitter—which over many itterations made these plants less problematic. After that, they processed these foods in multitude of ways to further reduce these compounds while also predigesting the seeds.

Our ancestors started to consume fermented and specifically

alcoholic foods well before agriculture. Primates evolved the gene to metabolize alcohol about ten million years ago so they could consume ripe fruit.[88] Chimps also consume fermenting fruit and saps that are about 5% alcohol—and appear to enjoy the effects.[89] Our ancestors probably continued these same practices, and *Homo erectus* possibly started to intentionally ferment some foods. After all, it merely happens whenever you add simple carbs to warm water inside any type of container. Despite this, scientists generally think that humans started fermenting alcohol only 9,000 years ago because that is when we have evidence from clay pots. But before then, our ancestors would have fermented foods inside animal skins or gourds that long perished. But during agriculture we greatly expanded our use of fermentation, including ones both high and low in alcohol. Agriculturalists usually ground, fermented, and cooked their grains, sometimes along with legumes, creating sourdough breads, porridges, or cakes. This neutralized most of the plant defenses, predigested the seeds into smaller pieces and added more nutrients while also improving the taste and texture.

Agriculturalists also fermented vegetables, fruits, and dairy—which also helped preserve these foods during the colder months, making them available year-round. They additionally soaked, sprouted, and roasted grains for similar reasons. Additionally, they also fermented grains, grapes, and various fruits into alcoholic beverages. To make beer they sprouted grains to break their starches into maltose, which is just two molecules of glucose. To make wines and cider, they mashed and strained the fruit into juice which was rich in sucrose. They then used wild yeasts and bacteria to ferment these sugars into alcohol as well as acetic acid, malic acid, lactic acid, and other organic acids, all of which give most fermented foods their sour taste. As noted before, these acids provide multiple benefits for our health and, in some cases, produce ATP.

Alcohol and organic acids require virtually no metabolism from our body to digest or assimilate. They are tiny, isolated compounds that flow straight through our body on the way to generating ATP.

Additionally, alcohol in particular generates more ATP per gram than glucose—so it probably produces more ATP in total than any other molecule (since fatty acids require considerable amounts of metabolism to utilize). For all those reasons, fermentation produced some of the most predigested foods ever—all from foods that, otherwise, are challenging to digestion.

So while agriculturalists consumed high amounts of grains, they did not consume them like moderns: they consumed these grains in some cases with the fiber intact and otherwise highly predigested and then fermented to include many additional nutrients, such as alcohol, various organic acids and additional nutrients like B vitamins, phytonutrients, antioxidants, etc. They consumed especially high amounts of acetic acid from three sources: from the fermentation of grains, from the fermentation of fiber in their colon (acetic acid is one of the three colonic fats) and from alcohol—which is converted into acetic acid inside our bodies.

Dairy Foods

In the past, foragers consumed small amounts of dairy from the stomachs and mammary glands of their prey. However, agriculturalists started regularly consuming larger amounts of dairy about 8,000 years ago and, in doing so, they provided themselves with one of the best sources of food possible—but one that also potentially came with some detriments or needs for adaptation. To acquire animal foods, humans usually kill the animal, but then it cannot produce any more protein. But with dairy they kept the animal alive to produce enormous amounts of protein, fat and carbs every day, along with every other nutrient humans need. Furthermore, the proteins are more complete than muscle meats and the fats somewhat balanced. Additionally, dairy is preservable through fermentation into yoghurt, kefir and cheese. So, in all those senses, dairy was one of the greatest foods ever.

But dairy is substantially different from other animal foods in several ways that are possibly problematic. This begs the question: to what extent did we evolve to consume this food? First, the fat in dairy contains unusual ratios of fatty acids, as compared to other animal and plant fats, with about 65% saturated fat, and only 25% or so monounsaturated and 3% polyunsaturated. Second, about 15% of the saturates are short chains that we usually only get from our colon, and medium chains that we get from coconuts. In other words, prior to dairy, humans never consumed fatty acids in these ratios before; and scientists do not know if we adapted, or if we needed to adapt, to this fat. Third, dairy contains lactose—that is, glucose bonded to galactose—that humans ordinarily cannot digest after the age of three. But many humans evolved to digest lactose about 7,500 years ago, while some did not.[90] Fourth, dairy is low in purines whereas all other protein foods, especially meat, organs and beans, are abundant in them. Purines are precursors to DNA, but in our body, they are converted into uric acid, which provides benefits as an antioxidant in our blood in the right amounts, but in excessive amounts these compounds lead to inflammation and eventually gout.[91] Again, scientists do not know if we adapted, or even needed to adapt, to consuming fewer purines in our diet. Finally, dairy contains enormous amounts of calcium, and given that agriculturalists probably consumed the equivalent of four to eight cups of dairy per day, they consumed enormous amounts of calcium, probably about three to ten times as much as their predecessors. Again, scientists do not know if we adapted to this. But once established dairy then became perhaps the primary protein of most agriculturalists for all the reasons mentioned above.

Foragers ate eggs but only occasionally and at certain times of year. But with the introduction of domesticated fowl, most of all chickens, agriculturalists made eggs one of the staples in their diet. Eggs are the most complete of all proteins, while containing about the best ratio of fatty acids for humans. Agriculturalists also started to use salt in greater amounts. As noted before, chimps and most other animals intentionally seek out salt and, presumably, our ancestors did as well, since we need salt for two of our primary electrolytes: sodium and chloride. However, agriculturalists started using salt to preserve foods and likely to either mask or enhance their flavor. So they likely consumed about two to ten times as much salt as their predecessors—which then causes the retention of water, which then leads to other detriments.

The Privileged and Unprivileged

Later on, many agriculturalists became parts of hierarchical civilizations like the Mesopotamians and Egyptians, as well as states like Rome and France where they were frequently separated into privileged and unprivileged classes. The privileged typically consumed plenty of animal proteins, both as completes and collagen, as well as animal fats.

They also consumed the grains and legumes of the day, as well as fruits, vegetables and alcohol. So in some cases, they consumed balanced diets but were also susceptible to excesses like moderns—usually an excess of sugar, salt, alcohol and meat.

However, the unprivileged, who were by far the bulk of the population, consumed almost entirely plant foods, mostly grains as bread, porridge and beer, sometimes with legumes to generate complete proteins but sometimes not. Generally, they could not afford or produce much, if any, animal foods because they did not have enough pasture, grain or legume to feed them. They tended to specialize in one or two grains as the bulk of their diet. In the morning, they drank sweet beer containing lots of maltose but not much alcohol—and then consumed beer with more alcohol throughout the day, so many lived most of their lives slightly inebriated—which helped them tolerate their lifestyle. Alcohol provided energy while also making their beverages safer to drink. At the same time, they consumed much lower amounts of fruit, berries, and vegetables so they suffered vitamin deficiencies, especially in vitamins A, C, K and possibly B12 as well as other forms of malnutrition. In other words, the bulk of our European ancestors lived in malnutrition and accordingly were short in stature, with males only about five feet two or four on average, with bad teeth, bad bones and other ailments as well as greater susceptibility to infectious diseases. Because of this disparity in diet, the privileged of course could more easily rule and exploit the unprivileged—who did not have the wherewithal to defend themselves.

IMAGE 14.2- FARMER

Herders

Some agriculturists known as herders, who emerged in Africa about 9,000 years ago, did not farm much, if at all, and instead mostly raised and consumed animals, including goats, sheep, cattle, pigs, camels and reindeer.[92] In some cases they lived almost entirely off their animals, such as the Mongolians and Sami, consuming both the dairy, flesh, and blood of their herds. In other cases they also consumed carbs that they gathered from farming, trading or foraging. When entirely carnivorous, they over-consumed protein at about 25% of total calories, fat at about 60%, and carbs at about 10 or 20% gotten from the lactose in milk. But they only consumed about 40 to 80 g per day of carbs, not enough for their needs, so presumably they lived in permanent gluconeogenesis and ketogenesis. They thus consumed nutrients in similar ratios to mostly carnivorous foragers, like the Inuit and Plains Indians. Despite veering away from the universal human diet, these people nonetheless exhibited excellent health and stamina, with the Mongolians in particular conquering much of the world that was more dependent upon grains and legumes.

IMAGE 14.3- MONGOLIANS

UNIVERSAL AGRICULTURAL DIET

Even amidst these four classifications of their diet, agriculturalists seemed to trend towards their version of the universal human diet whenever possible. About 20% of their calories came from protein, with 80% of that from completes (dairy, eggs, meat, organs and combinations of grains and legumes) and 20% from collagen (broth, skin, sausages, snouts, tails and the like). About 40 to 50% of their calories came from carbs, with at least 80% of that as starch from grains, legumes and tubers, which were properly processed and predigested and 20% of that as fructose from fresh fruit and honey along with many fermented foods, such as jams, wines and the like.

Because they consumed so many whole plants, they also generated considerable amounts of colonic fats. From culture to culture, agriculturalists of course varied in what plants and animals they domesticated—even as much of the underlying ratio of nutrients remained about the same.

FOODWAYS EQUATION OF HUMAN AGRICULTURALISTS COMPARED TO FORAGERS

The foodways equation applies best to animals, including humans, that live in the wild. However, I can still apply the equation to agriculturalists—in this case, the more privelaged of them.

CALORIES: 2000: decreased due to smaller stature and less exercise metabolism

=

MACRONUTRIENTS:

=

Protein: 20%: same as the percentage of calories.
Completes: 80% of protein: increased from dairy, eggs and combinations of grains and legumes.
Collagen: 20%: same.

+

Carbs: 50%: increased as the percentage of calories.
Starches: 80%: increased.
Fruits: 20%: decreased.
Alcohol, acetic and lactic acid: increased.

+

Fats: 30%: decreased due to abundance of grains.
Dietary: decreased for the same reason.
Colonic: increased from plant fibers.

=

METABOLISM: decreased

Sensing: same but not as necessary anymore for finding food.

+

Locomotion: decreased due to smaller territories and decreased foraging and hunting.

+

Capture: (hunting and gathering): decreased due to overcoming plant and animal defenses.

+

Ingestion: same.

+

Digestion: same, expensive.

+

Liver: (conversions and detoxification): decreased maybe, *expensive*.

+

Brain: (cognition and sociality): decreased in size, *expensive*.

Agriculturalists mostly reduced many of the obstacles from our foodways equation. Their food was not scattered through the wilderness, but located in smaller and captured territories. Instead of their animals running away or fighting back, they were captured and docile. They also reduced many of the defenses of plants through breeding. Agriculturalists also better protected their food from competition.

PERPETUATION OF AGRICULTURE

While agriculture did not necessarily improve, and even lowered, the quality of life for humans, it nonetheless proliferated across the world, becoming self-replicating like our genes and memes while destroying foragers and their wilderness along the way. This raises the question: why did agriculture spread so fiercely even while making humans less healthy? Agriculture propagated mostly because it allowed humans to breed more and increase the population. As noted before, wild animals are limited in their population by the constraints of food and territory, so human foragers could not increase their populations because they then could not feed them. However, farming and herding allowed humans to produce more food per area of land—and thus increase their population. Furthermore, once they had babies, human foragers then carried and suckled them for several years, which prevented ovulation—so they had only one baby per three years at the most. However agriculturalists weaned their babies after one year and then fed them porridge, so they were able to have many more babies. Even as many of these children died or suffered from malnutrition, they in the end greatly increased their population relative to foragers.[93] Once the population increased, agriculturalists could not then return to foraging, even if they wanted to, because the wilderness could not support their population anymore. By that time, too, they were so acculturated to farming that they did not have the health, strength and skills required for foraging.

Agriculturalists also propagated because they outcompeted foragers. Obviously, agriculturalists did not become superior to foragers as individuals in any way: they were shorter with slightly smaller brains and riddled with various forms of physical and social diseases most of the time. But agriculturalists could nonetheless outcompete foragers and conquer their lands because, as already explained, they were greater in population with bigger armies. While foragers lived in the open, easy to attack, agriculturalists protected themselves behind walls and fortresses that foragers could not overcome. While foragers were superior fighters as individuals, they could not organize themselves into disciplined armies, but agriculturalists could because they were conditioned to follow authority. While foragers tended to stall in cultural evolution, agriculturalists pursued ongoing invention, especially of tools and weapons. And finally, agriculturalists spread their infectious diseases to foragers and in some cases obliterated whole populations.

Interestingly, over the past few hundred years, many agriculturalists lived with foragers for one reason or another, and in most cases they never wanted to return to civilization. Meanwhile many foragers lived with agriculturalists but usually wanted to return to the wilds. All of this raises one of the more perplexing issues of our existence on earth: through cultural evolution, we humans invent and then replicate many memes that actually undermine the quality of our lives.

Once originating in the Levant, Agriculture spread in every direction, certainly northward into southern and then northern Europe, while not reaching Great Britain until about 6000 years ago. Along the way, the farmers and herders of the Levant interbred with the foragers in those areas, mixing their genes. Finally, the herders from Mongolia then conquered much of eastern Europe and interbred with those populations. But additionally, another group of herders from Russia and the Ukraine, known as Yamnaya, migrated into Europe as well. So modern people of European origins are mixtures of three types of foodways: sedentary farmers, nomadic foragers, and nomadic herders.[94]

CHAPTER 15

HOMO SAPIENS MODERNS

STORY

Starting somewhere around several hundred years ago, Homo sapiens transitioned from agriculture into what I call the modern. During this time most humans stopped producing their own food, both as individuals and communities, and relied instead upon large, centralized, and frequently exploitive organizations, like governments and corporations, to produce their food for them through the use of technology and strategies.

Along the way the motive behind our foodways changed. In the past our villages, families and ultimately our mothers and grandmothers controlled our diet, motivated by the desire to provide nourishing food to people they cared about the most. But now nameless people in corporations control our food, motivated by profit while manipulating our taste buds to eat malnourishing foods. At the same time, moderns abandoned many of our ancient culinary practices around selecting, combining, preparing, predigesting and presenting food and then replaced those practices with ones presented by capitalism. We now eat new foods and the usual foods in new and strange ways, including white flours, fruit juices, soda pops, vegetable oils, artificial ingredients, and increasing amounts of alcohol and other drugs—all of which we are not necessarily adapted to consume, especially in such amounts.

We also started to develop ideas about our diet, which ranged in their validity. Due to the changes in our diet, we started to experience even greater disruptions to our health and well-being — and then started developing theories about how to restore ourselves. Some of these were based on ethics: eat vegan, vegetarian, carnivorous. Some based on regions or ideologies: eat Mediterranean or Macrobiotic or Ayurvedic. Some based on eras: eat Paleo or traditional… and on and on. Meanwhile our scientists generated their own ideas, some good, some bad and while producing squillions of confusing and conflicting studies. But corporate marketers generated the central and most misleading idea: that we should eat enormous amounts of cheap, malnourishing, imbalanced and shelf-stable foods to make them money, not make ourselves healthy. So moderns not only do not eat our indigenous diet, but we do not even know our indigenous diet. However, while the modern created considerable detriments for our diet, it also created enormous benefits, at least in some senses — that is, we overcame scarcity, famine, and starvation while making all foods available year-round for little or no effort.

IMAGE 14.1- MODERN

ADAPTATIONS AND PATHOLOGIES

While our prior eras of evolution were initiated by changes to our ecosystems, the modern era was initiated entirely by cultural evolution. We did not adapt through natural selection at all because we have not passed through enough generations. Furthermore, selection is not even necessarily happening anymore, at least in the usual way, because outcompeting others is not required for mating. Additionally, we do not have much reason to believe that we adapted to the modern through inherent flexibility or epigenetics. Instead, we both created the modern entirely through cultural evolution and then adapted in the same way.

However, we did change in response to the modern—but those were not really adaptations. We changed our morphology by becoming taller than agriculturalists due to the abundance of food in childhood. Somewhere in their teens, twenties or beyond, most moderns start showing some evidence of the diseases of civilizaton due to their diet and lifestyle.[95] We are overweight to one degree or another and misshapen with tight muscles, stooped postures and bad teeth except for the intervention of dentists. When becoming overweight, we then start to trend towards many metabolic diseases, including insulin resistance, diabetes, heart disease and from there, we continue to trends towards many other afflictions not found in foragers, including autoimmune disorders, mental disorders, neurodegeneration, cancer, etc. With the help of sanitation and medicine, and distance from our livestock, we overcame many bacterial infections that plagued agriculturalists but, increasingly, we are susceptible to viral infections that do not have cures.

LIFESTYLE

Moderns generally live in enormous groups in towns or cities while not even knowing our neighbors or needing them for anything. In most cases we only form communities around isolated families, work, schools and in some cases churches. At the same time, both men and women do not perform the same jobs as their ancestors: men usually neither hunt, defend, herd, nor perform any physical labor at all. Women do not gather or farm plants or necessarily cook or tend the children — certainly to the extent as in the past. Furthermore, men and women are not divided in their labor anymore as they compete for the same jobs and cooperate in them as well.

We live in what I call facsimile culture: instead of experiencing our evolutionary lifestyle directly, we experience that lifestyle indirectly. Instead of eating real food, we eat fake food. Instead of nursing our children, we feed them formula. Instead of hunting, we play sports that are similar to hunting — or just watch them. Instead of patrolling our borders, we just watch movies or play video games about war and conflict. Instead of experiencing the dramas of life, we watch them in our movies. Instead of forming communities, we have friends and likes on Facebook. Instead of migrating from one camp to another, having adventures along the way, we take cruises and sip cocktails by the pool. Instead of having sex, we watch porn and on and on. Behind nearly every modern behavior is usually an attempt to express our evolutionary nature, albeit in direct and usually weird and less fulfilling ways.

As explored in my other book, *On the Origin of Being*, we humans now experience a considerable mismatch between our evolutionary selves and our cultural selves. Since we are not biologically adapted to our cultural evolution at this point, we experience mass pathology in our bodies not only because of our diet and lifestyle but also because of the way we sleep, wake, rest, work, sit, and isolate ourselves from nature.[96] We also experience pathology in our families, societies, and cultures for the same reason.

DIET

During the modern we overcame nearly all obstacles to attaining our food, including scarcity, defenses, competition, and the like—which in turn made food abundant, convenient, and cheap. We can now attain food with little, if any, effort at all, so we now overeat more food than we need. We changed the way we consume protein, over-eating too many complete proteins as eggs, dairy, and meat while eschewing collagen proteins—even as we do not usually combine our grains with legumes and nuts. We also consume protein powders—all with unknown consequences to our health. Through this excess and imbalance in our protein, we utilize less of the protein we consume and convert more of that protein to ketones and glucose while generating waste products. At the same time, we do not anabolize our collagenous tissues well enough, leading to pathologies in those tissues, including problems with joints, skin, teeth, bone, hair and fascia and probably even our cardiovascular and nervous system. So overall we eat lower-quality proteins than our ancestors. We also eschewed organ meats, especially liver, depriving ourselves of one of the few sources of preformed Vitamin A.

We started to over-process our carbohydrates. As noted before, we evolved to select for more refined foods and then process them for our benefit. But during the modern we processed our foods by stripping away many of the nutrients—all to save on the cost of production and extend shelf life. We removed the bran and germ from our grains to make white flours. We removed the skin, pulp and seeds from our fruits to make juices. And we refined several plants, sugar cane, beets and corn, to render pure sucrose. Through this form of processing, we reduced the antinutrients and toxins, which tend to congregate in brans and skins, while making these foods easier to digest and more abundant in macronutrients, so in those ways, we followed our instincts and even provided benefits to ourselves. But paradoxically, we also removed many nutrients, including minerals, vitamins, phytonutrients, essential fatty acids, as well as all the additional nutrients and benefits from fiber—all of which resulted in greater detriments that trended us towards the diseases of civilization in ways already explained.

Moderns also consume more fats, at least as compared to most agriculturalists. We generated enormous amounts of animal fats from feedlots where animals are almost force-fed to make them fat. For the first time in our existence, we started to both produce and consume enormous amounts of plant oils—and in turn, altered the way we consumed our ratio of fatty acids—that is, less saturates and more monounsaturates and polyunsaturates. Additionally, when polyunsaturates are consumed in such high amounts in such refined forms, they possibly generate considerable detriments related to eicosanoids as well as compounds called OXLAMS.[97]

Moderns also continued to add more salt to our diet. We also introduced artificial flavorings, colorings, and preservatives and many other compounds—all of which are generally regarded as safe but not necessarily safe, especially considering the amounts many consume. In many cases these compounds, when used as flavorings, mask the bad taste of non-nutritious foods, essentially tricking our taste buds into eating them.

We also over-consume salt, and possibly calcium, relative to other macrominerals and under-consume magnesium. Even more, most moderns now consume multiple supplements—even though nearly all these supplements are not proven as beneficial and are possibly even detrimental. Additionally, we consume more and more pharmaceutical and recreational drugs to mask our symptoms. Possibly worst of all, we stopped balancing the nutrients in our food in many cases, especially with snacking, consuming foods high in carbs, plant oils and salt. As explained in prior sections, this induces our body to overeat carbs and fat and undereat protein, resulting in weight gain.

FOODWAYS EQUATION OF HUMAN MODERNS COMPARED TO AGRICULTURALISTS

In the modern, the foodways equation becomes increasingly not applicable; as noted before, the equation really applies to animals, including humans, living entirely within their biological evolution—that is, in the wilderness—while we are now living within our cultural evolution. But I will nonetheless outline the equation for moderns, which does indeed elucidate our condition.

On one side of the foodways equation, we radically raised our consumption of calories for all the reasons already mentioned. On the other side of the equation, we radically lowered our metabolism—resulting in excess calories that essentially converted into fat and trended us towards the diseases of civilization. We lowered our basal metabolism; our digestive metabolism and our exercise metabolism. While our ancestors catabolized about 20 to 40% of their calories for exercise metabolism, moderns catabolize about 5%.

CALORIES: increased: about 2500 calories per day due to the excess of food and eating for recreation.

=

MACRONUTRIENTS: increased refined foods but removed nutrients

=

Protein: same: 20% of overall calories even though we do not need this amount of protein.
Completes: (decreased).
Collagen: (decreased).

+

Carbs: increased: 50% of overall calories.
Starch: increased.
Fruit: decreased.
Sugar: increased.

Note: eat more carbs than ever before, most of them as over-processed, and more fructose relative to glucose.

+

Fats: same or decreased: 30% of overall calories.
Dietary fats: increased: 28%.
Colonic fats: decreased to about nil in some cases: 2%.
Note: greater amounts of monounsaturates and polyunsaturates relative to saturates.

=

METABOLISM: decreased radically, both basal, digestive and active.
Sensing: same.

+

Locomotion: decreased to nearly insignificant.

+

Capture: (hunting and gathering): decreased to nearly insignificant.

+

Ingestion: decreased due to over refined foods.

+

Digestion: decreased: eat more food but easier to digest, *expensive*.

+

Liver: (conversions, storage and detoxification): increased conversions due to lipogogenesis and imbalanced diet; decreased detoxification due to refined diet.

+

Brain: cognition and sociality: same, *expensive*.

Our foodways equation is not in balance. We radically increased our overall consumption of calories while radically decreasing our level of metabolism for basal, digestive, and exercise. We eat too much of all macronutrients and eat them in the wrong ratios, with our proteins and fats imbalanced and our carbs toxic while eating little fiber, fruits, vegetables, phytonutrients or liver with their abundance of key nutrients. Even though all of this is easily fixable, we nonetheless choose to persist in this condition and live onward in our own form of submission and pathology, called the diseases of civilization which, when all is considered, is both physical, social and psychological—and even spiritual.

CHAPTER 16
SUMMARY AND CONCLUSION

To survive, all animals, including humans, must have foodways—that is, they must acquire food and utilize the underlying nutrients. This leads to the foodways equation. On one side of the equation is calories—the sum of the macronutrients in the diet. On the other side is metabolism—the steps to acquire those calories, including sensing, locomotion, ingestion and digestion, as well as the steps to utilize those nutrients, including metabolism, conversion and storage. We also use most of our brain in this metabolism, including the autonomic and cognitive functions. During their foodways, animals encounter obstacles because their diet is distant, scattered, defended, and under competition from other organisms. To overcome these obstacles, animals evolve not only foodways but precise and efficient foodways, making them magnificent creatures.

All animals evolved specific foodways for their ecosystems over time—so that their foodways equation is balanced. When animals acquire more nutrients than needed for their metabolism, they tend to breed more; and when acquiring less, they breed less. Even then they sometimes still cannot acquire enough nutrients so they must adapt through inherent flexibility, epigenetics, cultural evolution, and natural and sexual selection to restore balance to their foodways equation. Overall, these dynamics controlled much of the evolution of all animals, including humans, as we have seen.

CONCLUSIONS

If we compare the premises of the optimum diet, as outlined at the beginning of this book, to the premises of ancestral consumption, we see that they are similar at least all the way through foragers and then even some to agriculturalists and moderns.

Premise 1 is do no harm. Our ancestors did not eat over-processed or fake foods, additives, drugs, or supplements because they were not available, and we therefore did not evolve to consume them. However, our ancestors did consume some substances that perhaps are defined as supplements, including clay, mud, salt, and medicinal plants.

Premise 2 is optimize digestion, or said in the converse, reduce digestive metabolism. Obviously, we reduced digestive metabolism throughout our evolution and all the way to the present through choosing more refined foods, and then processing and predigesting those foods. So we use less of our calories for digestive metabolism while using more for other forms of metabolism. This works in humans every day and, in the past, drove our evolution: through needing fewer calories for digestive metabolism, we provided more calories for our brain and other metabolisms.

Premise 3 is reduce plant defenses, including antinutrients and toxins. Our ancestors did this through selecting refined foods in the first place, especially fruit and animals. Later they reduced these compounds through soaking, fermentation, and cooking; and during agriculture, cultivated plants to contain less of these compounds. During the modern we continued this trend but, in the process, stripped away too many nutrients, leading to more harm than good. Through reducing these compounds, our ancestors were enabled to shrink their livers and reduce the effects of these compounds on their own physiology.

Premise 4 is to consume all nutrients. In every case with every ancestor, we see them eating all nutrients whenever they could acquire them in their ecosystem — and not eschewing some nutrients in favor of others, or deeming some good or bad. They consumed complete proteins from animals and later from dairy, eggs and combinations of grains and legumes. They consumed collagen proteins from animals. They consumed carbs as fruits, tubers, grains with their fibers intact. They consumed all forms of plant and animal fats which tended to balance them in certain ways. They consumed liver for preformed vitamin A and other nutrients, as well as colorful fruits, leaves, and herbs for various vitamins and phytonutrients not found in the remainder of their diet. However, humans can obviously survive and even thrive without eating enough of all nutrients, as witnessed by the extreme carnivorous humans like the Inuit and various herders like the Mongolians — who might have various genetic adaptations for this diet. Even these these people tended to focus on animal foods that contained carbs like dairy.

Premise 5 is that they ate the optimum number of calories per day. Generally they lived in balance with the foodways equation so that calories equaled metabolism.

Premise 6 is they consumed all nutrients in their proper ratios — again, whenever they were available. In all cases, our ancestors did not gorge on some nutrients at the expense of others but consumed all of them in balanced ratios. Ultimately, these proper ratios accord with the approximate specifications of the Universal Human Diet and with nutritional physiology.

Interestingly, balancing taste does not apply to all our ancestors. Our theoretical ancestor and *Australopithecus* consumed their foods hand to mouth — so they were not balancing flavors, but just eating on the spot whatever they hunted or gathered. However, with *Homo erectus* and beyond, our ancestors carried their foods back to their hearth where they combined them in ways to balance their flavors between sweet, salty, umami, and acidic while reducing bitters and astringency—which in turn helped them to optimize their nutrition as each taste correlates with specific nutrients.

In every way, this ancestral consumption accords with what we learned about nutritional physiology, bringing my theory further to the truth.

Obviously, during the modern, we altered these premises.

Premise 1: We did away with this one entirely and bombard our bodies constantly with all sorts of weird and potentially damaging compounds, ranging from altered foods to pharmaceutical drugs and fancy supplements.

Premise 2: We continued to optimize digestion, but to the extreme. In over-processing our foods, we over-consumed them and deprived ourselves of nutrients, trending us towards the diseases of civilization.

Premise 3: We continued to reduce plant defenses but, in the process, removed many nutrients.

Premise 4: We are not especially good at eating all nutrients, even though all of them are available because we regularly eschew many vitamins associated with plants, including vitamin A, C, K and E—as well as collagen and preformed vitamin A.

Premise 5: We live in imbalance with the foodways equation. We regularly over-consume calories and then under metabolize them, especially by reducing our exercise metabolism.

Premise 6: We are not especially good at eating all nutrients in their proper ratios, as we regularly consume foods enormously high in sugar, sometimes in combination with fat, but low in protein and nearly all other nutrients. We regularly do not eat complete proteins, nor collagen. We over-consume carbohydrates, especially refined ones as white flours, sugars, juices and the like. For the first time in our existence, we consume enormous amounts of polyunsaturates and monounsaturates from plant oils while under-consuming satu-rates—all while consuming imbalanced ratios of omega-6 to -3. We also do not consume our macrominerals in the best ratios, with usu-ally too much sodium, chloride and calcium and too little potassium and magnesium.

Premise 7: We do not necessarily balance our tastes all that well, either, at least in ways to signify the proper balance of nutrients. Instead, we use artificial ingredients and other substances to mask and alter our tastes.

While we can learn enormous amounts from ancestral consumption, we cannot necessarily just eat like our ancestors because we do not live like them and, most particularly, we do not exercise nearly as much as them—which in turn affects the amount of protein and fat needed in our diet. We need to understand ancestral consumption and then integrate that understanding into our own lifestyle and diet—as I will describe in the upcoming section.

In this story of ancestral consumption, we see the larger premise of this whole book—to optimize our diet. While evolution is not necessarily moral, it nonetheless designs us to survive, thrive, compete, cooperate and reproduce—and thus optimize our biological potential. Accordingly, evolution and natural selection in particular design us to consume certain foods in certain ways to optimize our biology. So in learning the diet of our ancestors, we largely learn the diet that optimizes our biological potential. I did not contrive the premise of this book—to design our diet to optimize our health. Instead, this premise is one of the fundamental and most relevant drivers and truths of our existence—the biological imperative.

PART 3

THE PREMISES

In the prior sections, I covered two of the four methodologies to determine our optimum diet: nutritional physiology and ancestral consumption. Before I cover my seven premises in depth, I will briefly discuss the other two methodologies: establishment recommendations and scientific studies.

Establishment recommendations cover the experts, the various scientific and governmental organizations, including The Food and Nutrition Board of America, the National Academy of the Sciences, the European Food Safety Authority and the World Health Organization. These organizations are overseen by people with academic credentials on some specific and specialized topic like carbs or certain vitamins. Most of these organizations make recommendations, DRI (dietary reference intake) or the RDA (recommended daily allowance) for our daily intake of calories, macronutrients and most other nutrients as well, while also setting their lower and upper intake levels. They also monitor the nutritional panels on the back of our packaged foods. Generally, these organizations tend to reach the same or similar conclusions and recommendations so, for that reason, we should listen to them.

However, we also should exercise some skepticism about them. These organizations, first of all, are not attempting to define the optimum diet but only the adequate diet to avoid malnutrition and disease, so you cannot necessarily attain optimum health by following their recommendations. Second, they are possibly hamstrung by the specialized and narrow training of their experts; perhaps they are experts on one class or even type of nutrient while not understanding the larger systems involved in nutrition. While mostly relying upon scientific studies for their determinations, they tend to ignore both nutritional physiology, at least to some extent, as well as "ancestral consumption"—thus relying mostly upon one methodology.

Third, organizations like this, even though scientific, nonetheless trend towards letting politics and economics interfere with the truth. They also trend towards being dogmatic, stuck in ideological conformity and groupthink, while avoidant of risk and complexity and slow to change—and even slower to address past mistakes while ignoring present ones altogether. Since they need to present unified presentations to the public about nutrition, they will trend towards only selecting conformists with the best credentials — while avoiding independent researchers following their own instincts. Even as these establishments have existed since the early 1940s and while possibly providing some or even great benefits for humans to better their diet, Americans have become increasingly diseased by their diet. This, perhaps, is due to Americans ignoring their recommendations — but maybe that is because some of these recommendations are unappealing and wrong. While getting most things right, the establishment nonetheless gets some things wrong — even ones they now admit as wrong. In the end, given the positives and negatives, I always strongly consider and respect establishment recommendations but exercise skepticism about some of them.

My final methodology, scientific studies, are studies published in reputable and peer-reviewed journals — even though the findings are not necessarily part of the establishment for various good and possibly bad reasons. These studies come in various types. In observational and epidemiological studies, scientists track the relationship between diet and health and disease, in groups and even whole populations from the past and present. Longitudinal studies are similar except they tend to focus on groups within larger populations and tend to collect data into the future. Cross-sectional studies collect data at one point in time — and look for correlations. Nutritional surveys are also similar but track data using surveys, usually given to either large or smaller populations. While quite helpful, these studies are nonetheless limited for several reasons. First of all, like establishment recommendations, they are mostly concerned with disease prevention, not optimum diets. Second, these studies focus on whole groups of people who perhaps eat similar but not the same diets — so the controls are somewhat lax. Scientists attempt to smooth out these problems with various analytical tools, somewhat successfully, but in the end, these studies are just suggestive not conclusive.

In clinical trials, and specifically double-blind studies, scientists divide subjects into two groups and then feed one group placebos and the other group one nutrient or supplement. Additionally, the scientists do not know which people are in which group. Afterwards scientists compare the placebo group to the substance group, and track correlations between that substance and health. These studies, however, have their own limitations because they usually track only one nutrient in isolation while, in reality, we ingest dozens at once through our food. Given this, we cannot necessarily assume that one nutrient will have much, if any, measurable or significant impact upon our health.

Obviously, in this book, we are seeing the interconnections between all nutrients. At the same time these studies never consider two of the most confounding issues in nutrition—that is, nutrigenetics and epigenetics. These studies just assume, more or less, that all humans are mostly the same, and that one study of some humans should apply to all humans. With that being said, I consider scientific studies quite helpful, but exercise skepticism about them.

Since all four of my methodologies reveal truths while containing limitations, I do not rely on one but all my methodologies—and then compare them to each other, while looking for the ways they corroborate and contradict each other—and then attempt to resolve the contradiction before arriving at my final premises and recommendations. So in this section I will analyze my seven premises with the four methodologies—and then make my recommendations for optimizing your diet for your lifestyle. In the final section of this book, I summarize all these premises for your easy reference.

CHAPTER 17

PREMISE 1 – DO NO HARM

"Do no harm," of course, is taken from the Hippocratic Oath. Or said in more precise terms, exercise caution in consuming supplements or foods not indigenous to our diet, including artificial ingredients, vitamins, minerals, protein powders, etc. While generally regarded as harmless, these substances possibly give benefits but also cause detriments in ways that are not clearly detectable.

NUTRITIONAL PHYSIOLOGY

We moderns are conditioned to think we can solve many of our issues with technology. However, as we have seen, our nutritional physiology evolved over many millions of years and has not changed much in the past ten thousand years and not at all over the past several hundred. So, this raises the question: is our ancient design compatible with these new technologies, with these pills and powders? If our bodies are designed to extract vitamins and minerals from our food, should we now receive these substances from pills? If our bodies are designed to extract amino acids from muscle, broth and legumes, should we now receive those amino acids from protein powders? Should we not even consider the possibility that these types of things worsen our health over the long haul? For this reason, I recommend that we avoid any supplements or foods our ancestors did not consume before two hundred years ago — or exercise caution about them and only consume them in small amounts for specific reasons or outcomes.

ARTIFICIAL AND ORGANIC

In the sections below, I will apply the principal of "do no harm," to various parts of our food and supplements, including artificial and organic ingredients, vitamins, minerals and then protein powders and others. Along the way, I will adress the four methodologies while making recommendations which, as mentioned, are summarized in the final chapters of this book.

First, you should generally avoid all artificial ingredients and preservatives. Second, you should eat organic plants when- ever possible, like our ancestors prior to hundreds of years ago, as they are less likely to contain artificial toxins, like insecticides, that possibly harm our physiology. Third, scientific studies show that these organic foods tend to contain greater amounts of nutrients.[105] Since these plants must defend themselves against predators, instead of relying upon added chemicals, they are essentially stronger and more nutritious — which probably passes benefits onto us. Presum- ably, the nutritional establishment would recommend organic as well, except they cannot risk alienating conventional farm- ers who supply the bulk of our food.

Fourth, you should eat organic animals as well — or animals raised as closely as possible to their natural conditions, including pastured chickens and grass-fed ruminants. When these animals are fed commercialized diets in alien environments, like feedlots, cages and pens, they are more likely to diseased in some way and generally need antibiotics and other chemicals to keep them alive — and this possibly carries into their flesh in ways that are not understood.

VITAMINS

Fifth, you should generally avoid vitamins and supplements: if eating balanced or optimum diets, you do not need them. Our ancestors stopped making vitamins inside their bodies many eons ago because they are so ubiquitous and attainable in our diet; additionally we can store enough in our liver to last weeks and months.[106] Our ancestors did not take vitamins. Although setting "upper limits" for them to avoid toxicity, the establishment does not recommend vitamins because they cannot prove any benefits either as multi or individual vitamins—and but they sometimes prove detriments.[107] However, if vegan, you may want to consider some possible exceptions to this rule, because when not eating animal foods, you likely deprive yourself of certain nutrients, most of all B12 and possibly B6, iron, as well as Vitamin A (as some people cannot convert enough of the carotenes into preformed vitamin A).[108, 109]

MINERALS

Sixth, you should also avoid most forms of mineral supplements as well, provided you are eating balanced diets. We are designed to extract macrominerals from both our small and large intestines. When eating more minerals than we need, we in some cases uptake them anyway and then excrete them in our urine. When not eating enough, we tend to uptake more of them and, instead of excreting them, recycle them back into our bloodstream. And since they are atoms, they never break down and thus we can reuse them indefinitely. So we have systems in place to regulate these substances even under difficult situations.

In most cases, our ancestors did not take mineral supple- ments with some exceptions. Again, the nutritional establishment recommends that you receive your minerals from your diet. Scientific studies show benefits to supplementing with some minerals but assumably, only when consuming unbalanced and over-processed diets where much of the magnesium and potassium are stripped away.[110]

However, there are some exceptions to this rule: salt, iodine, and clays (which are collections of minerals). We use salt, specifically sodium and chloride, as two out of the six electrolytes that are most abundant in our blood. While animals evolved in the ocean, using salt as one of their electrolytes, many animals transitioned to the land where salt is scarce in our plants, animals and waters. So most land animals, including primates, intentionally seek salt from mineral deposits, animal blood and specific plants. The nutritional establishment recommends some amounts of salt. And scientific studies show that salt in the right amounts provides benefits, but in excess causes detriments.[111,112]

We use iodine to make thyroid hormones. However, since some ecosystems do not contain enough iodine in their soil, animals must specifically supplement this mineral, including humans who, for example, specifically travel long distances to get seaweeds rich in iodine.[113,114] More recently, many Americans suffered from iodine deficiencies, developing goiters until iodine was added to our salt. So you should supplement with iodine with seafoods and seaweeds or iodized salt. Many animals, including our ancestors and even modern humans, use clay or charcoal to help detoxify their guts—as clay binds with toxins and parasites and helps eliminate them through our stool.

PROTEIN POWDERS

Seventh, I recommend that you not consume protein powders but receive your protein from whole foods, including plants and animals, because our nutritional physiology is not designed for these powders. Our ancestors did not consume these powders—even though they did turn animal tissues into powders and liquids, as with pemmican and broths. The nutritional establishment recommends that you get your protein from whole foods. Scientific studies, however, are somewhat mixed, suggesting some detriments while also some possible benefits.

These powders are potentially detrimental because they are over-processed—that is, they strip away many of the additional nutrients, including fat, carbs, vitamins, minerals, nucleotides, purines and phytonutrients, so in that sense, they are like table sugar. While they probably do not cause damage in the short term, these powders likely cause deficiencies in some nutrients in the long term, because these powders are so compact, we are inclined to overeat them and then digest them too quickly both of which possibly sends more amino acids into our blood faster than we can utilize them, forcing us to convert them into glucose and ketones. Additionally, in the process of making these powders, using in some cases extreme heat and pressure, some of the amino acids are potentially damaged, especially the ones that contain sulfur.[117]

However, there are some possible exceptions to this rule. Some studies suggest that whey protein powder derived from milk is best for sports recovery and muscle growth, because this protein absorbs quicker and thus works faster while also possessing certain amino acids that enhance muscle growth.[118] However, whey protein is also found in milk, at about 20% of the overall protein, with the remainder as casein. But because whey absorbs faster and casein slower, milk thus provides more sustained protein for anabolism.[119] Studies show that the protein powders derived from collagen provide considerable benefits. These come in two forms: gelatin, which are proteins, and peptides, which are those same proteins reduced into smaller chains. Because they are predigested, these peptides supposedly digest faster and work better.[120] While scientists have proven that these peptides provide advantages, they have not proven that they work better than gelatin or broth or other natural forms, such as animal skins. However, some collagen powders are whole proteins from whole foods, just in dried form.

So the question remains: why not just consume eggs instead of egg protein? With that egg, you get all the fats in the yolk as well as many other nutrients. Why not just consume dairy instead of whey protein unless you cannot digest lactose? Why not just consume peas instead of pea protein?

CHAPTER 18

PREMISE 2 – OPTIMIZE DIGESTION

This diet recommends that we prepare our food for optimum digestion—for complete and efficient digestion. Complete digestion means that our food is broken down into nutrients and then absorbed into the blood, to prevent that food from flowing into our colon and producing various detriments. Efficient digestion means that our food reduces "digestive metabolism" so we use more of the calories in our food for other forms of metabolism.[98,99]

NUTRITIONAL PHYSIOLOGY

In this book I define digestion as the process of breaking down our food into smaller particles, then into macronutrients and finally into individual nutrients like glucose—which are then absorbed into our blood and transported to our cells. I divide digestion into two categories: one called pre-digestion, which happens before we put our food into our mouth; and digestion, which happens after.

Pre-digestion

We humans start "pre-digestion" when cutting, mashing or grinding our food into smaller particles — such as flours — or soaking, sprouting and fermenting as mentioned. We then almost always cook our food, which causes proteins to denature, meaning their three-dimensional shapes are flattened, so that the links between amino acids are more exposed to digestive chemicals. Carbohydrates are broken down into smaller carbs, softened, gelatinized and engorged with water, so that the remaining bonds are more easily severed.[100] Fats, however, are not changed much during most forms of cooking but during prolonged cooking, possibly divide into fatty acids and glycerol.[101] So even before putting food in our mouths, we have already predigested our food to some extent, but pre-digestion does not render individual nutrients like glucose, amino acids and fatty acids in almost all cases.

Digestion

We start digestion by putting food into our mouths and chewing that food into smaller particles while coating it with saliva. Carbs, too, are coated with the enzyme, amylase, that breaks down carbs into smaller chains. After swallowing food into our stomach, we then release hydrochloric acid, which further breaks down our foods and activates various enzymes, especially pepsin, which digests protein while also killing any pathogens. Once dumped into our small intestine, the macronutrients are then broken down into their smaller compounds — i.e. protein into the twenty-one amino acids, carbs into the three sugars, and fats into the various fatty acids. Then all those nutrients, along with minerals, vitamins and phytonutrients, are taken up into our blood for transport to our cells for metabolism, conversion or storage.

Interestingly, our food undergoes an enormous number of steps, about fifty to one hundred, both outside and inside our bodies, before finally performing its ultimate purpose: the production of ATP and proteins and other molecules. Let's consider turning whole wheat into ATP. During pre-digestion we grind the grains into smaller particles, and then ferment and bake them. During digestion we chew the bread, breaking it into smaller particles, while amylose starts to break down the larger carbs. Once in our small intestine, the various enzymes continue to break down the carbs, finally into the sugars. Through several steps we then transport that glucose through our blood into our cells where glucose is taken up into the cells to undergo about ten steps of Glycolysis to render ATP, as well as two molecules of pyruvate, which then enter into the Krebs cycle to undergo about another ten steps to generate ATP, and then further steps are taken through the electron transport chain.

Complete digestion

If not digesting our food as completely as possible, we generate detriments for ourselves. We never digest all of our food: trace amounts of proteins and fats always dump into our colon and are then utilized by our own microorganisms. However, we do not ever want more than those trace amounts in our colon (outside of fiber and resistant starches) because they become toxic. Excess proteins ferment in our colons into ammonia, amines, phenols and sulfides which are potentially detrimental.[102] Excess fats in the colon do not ferment, as microorganisms do not usually ferment fat, but nonetheless cause bloating, discomfort, incontinency and urgency.[103] Fructose in the colon ferments into hydrogen and methane gases, which causes pain, bloating, flatulence and diarrhea.[104] For that reason, we want to digest our food as completely as possible.

Efficient Digestion

Also, if not digesting our food as efficiently as possible, we increase our "digestive metabolism," otherwise known as the "thermogenic effects of food." Scientists calculate digestive metabolism as somewhere around 8 to 20% of our "TEE" (total energy expenditure). They also calculate it as the percentage of calories in the food needed to digest the food. Or in other words, if one piece of bread contains one hundred calories, and you need ten calories to digest that food, then that bread requires 10% of its calories for digestion, leaving only 90 calories for other metabolisms. Protein requires between 20 to 30% of its own calories to digest, while carbs are 5 to 10%, and fats are even lower at 0 to 3%.[105] Interestingly, soluble fiber is at 0% because we use the digestion of the microorganisms in our colons. However, most vegetables, especially when raw, probably require about 50% or more of their calories to digest for two reasons: they do not contain many calories and contain compounds that impede digestion.

ANCESTRAL CONSUMPTION

Most animals prefer to optimize their digestion—that is, to digest their food as completely and efficiently as possible—by preferring to eat the most refined foods within their natural diet and even some outside of their natural diet. When given the choice, mice prefer pounded and cooked food, and especially cooked food, to raw and whole food.[106] Various primates showed the same preference as evidenced by several studies.[107] Occasionally chimps seek cooked foods in the aftermath of wildfires.[108] So animals evolved to optimize their digestion by selecting the most refined foods available to them, within the constraints of their design.

Our own ancestors optimized their digestion in several ways. First, our theoretical ancestor selected more refined foods and chewed their food for about five hours every day.

Later, *Homo Erectus* used tools to prechew and predigest their food through mashing and grinding. Agriculturalists amplified pre-digestion through fermentation, even making drinks with alcohol and acetic acid that did not require any digestion. Also we evolved towards consuming more fat, as rendered oils and tallows, which as mentioned require the least amount of energy to digest. Modernists then exaggerated these trends, into detriments, by rendering foods into single nutrients like sugar and protein. Presumably, we have the lowest digestive metabolism of nearly every animal on earth, with some exceptions, such as carnivores that specialize on fatty animals.

ESTABLISHMENT RECOMMENDATIONS AND SCIENTIFIC STUDIES

The establishment does not make any recommendations about optimizing digestion at all. But they recommend that we consume whole grains, beans, nuts and vegetables without any guidance on processing them—so they are mistaken. At the same time, they recommend that we avoid over-processed foods—even while most do not follow this advice.

Scientific studies have proven the effects of pre-digestion on food. During sprouting grains use their own enzymes to break down larger nutrients into smaller ones.[109] During fermentation microorganisms do the same, so that in both cases proteins break down into smaller chains of amino acids; larger carbs into smaller carbs; and in some cases triglycerides are broken down into fatty acids and glycerol.[110] Fermentation, if prolonged, will even break down gluten into amino acids. At the same time, these processes enhance vitamins and other nutrients, like acetic acid, while fermentation in particular adds more flavor and nutrition, mostly from organic acids. When consuming properly fermented whole grains, humans show improvements in digestion and gut health, the regulation of blood sugar and a reduction in inflammation.[111] Paradoxically, some studies attempt to show that through increasing digestive metabolism, humans can lose weight because they use more calories to digest their food and thus store less as fat. However, thus far, this theory has not been proven to work.[112]

So in this case, three of our methodologies are in accord with each other—that is, that we should properly process our food to optimize digestion. However, the nutritional establishment does not agree with this—and in this case, I think they are just wrong and probably beholden to lobbyists whose companies do not want to process their foods in the proper ways.

RECOMENDATIONS

So you should optimize your digestion by making your foods more completely digested to keep them from entering your colon; and more efficiently digested to reduce our digestive metabolism. To optimize digestion, you should first select more refined foods: animal foods, tubers, fruits, grains, legumes, etc., while eating vegetables in moderation. And then properly predigest your foods, using many of the same techniques used to remove their secondary compounds.

You should grind, soak, sprout, or ferment (or some combination of those) your whole grains—and then thoroughly cook all of them. You should soak your beans in either water or acidic water, and possibly let them germinate—and then thoroughly boil them. Beans generally do not ferment well except as tempeh.

You should soak nuts overnight in water or possibly acidic water or brine—and then dehydrate them or grind them into porridges or pancakes. Once soaked you usually cannot roast them as they burn easily.

You should consume all fruits either raw or cooked. While vegetables are recommended, especially because they contain nutrients not found in other foods, they nonetheless contain fewer calories than are needed to digest them, so we should consume them in moderation, especially when raw, and consider pureeing and cooking them.

Generally, all meats are completely digested even when raw. However raw meats require considerably more digestive metabolism as well as time.[113] So you should consume foods like oysters and sushi, and possibly rare steaks, in moderation and otherwise cook your meats. During cooking, most of the protein in muscles and organs start to denature at lower temperatures (at about 140 degrees), but higher temperatures expedite the process. However, collagen does not start to denature and dissolve in water until around 180 degrees—which then makes collagen much easier to digest for humans.[114] You can predigest animal protein in various other ways by grinding tissues into sausages and burgers; by ageing meat so that enzymes and microorganisms predigest the proteins and fats; by curing meat whereby the salt denatures proteins; by marinating meats in acidic mediums; as with ceviche, which denatures proteins as well.

When fats are lodged into foods, such as nuts or meats, you should then predigest those foods in the usual ways. But when fats are rendered into butter and various oils, we cannot break them down into smaller molecules through cooking—but in any case, fats digest more completely and efficiently than any other food.

Cooking predigests our food but also potentially damages the nutrients in our food. So you should strive to optimize pre-digestion while minimizing the damage to nutrients through controlling the temperature, time and method. Generally, all foods and nutrients start to predigest at around 140 degrees while collagen requires 180 degrees as noted before. Usually lower heat cooks more slowly; higher heat, faster. Based on my review of various studies, it seems that cooking foods at or under boiling, around 212 degrees, does not appear to cause damage to any nutrients, even with prolonged cooking times, such as one to two hours, with the exception of vitamin C, beta carotene and some B vitamins.[115] However, higher temperatures, especially ones in the 300 to 500 degree range, as found with grilling, broiling, baking and frying, tend to damage some or even all nutrients, especially as the time increases, including amino acids, sugars, fatty acids and micronutrients. Many foods, packaged in cans, boxes and pouches, are cooked at over 300 degrees under high pressure, the process known as retort, which damages nutrients while also destroying color and taste molecules and completely sterilizing the food.[116]

I could argue that we humans should reduce all of our food into flours, porridges, purees and then thoroughly cook all of them—and it does seem that our ancestors, especially agriculturalists, ate most of their foods in this way, as pemmicans, breads, porridges, jams, sausages, broths, beers, wines and the like. But we moderns appear to like some texture, chewiness and crunch to our food. But when eating foods in larger particles like this, you should chew these foods thoroughly before swallowing.

CHAPTER 19

PREMISE 3 – REDUCE TOXINS AND ANTINUTRIENTS

NUTRITIONAL PHYSIOLOGY

The optimum diet recommends reducing antinutrients and toxins in our plant foods. As we know, plants are designed by evolution, the same as us, to survive and reproduce—and thus to avoid being eaten into extinction. So they cover themselves in mechanical defenses like barks, husks and thorns while also growing in places that are hard to reach, such as under the ground or high in the sky. But plants mostly defend themselves with chemical defenses, known as "secondary plant compounds," both antinutrients and toxins.

Antinutrients block the absorption of nutrients inside our guts,[117] with these compounds in particular blocking protein: tannins found in fruits, especially grapes and many seeds; oxalates, trypsin and protease inhibitors found in leafy greens, whole grains, nuts, berries, tea, etc; and protease inhibitors found in most grains, as well as potatoes and some legumes. These compounds block the absorption of carbohydrates: amylase inhibitors found in most seeds; lectins found in fruits, seeds and many other plant foods. Phytic acid prevents the absorption of phosphorous, calcium, magnesium, iron and zinc and are abundant in all seeds, including grains, beans and nuts.[118]

Plant toxins poison various tissues inside our bodies, like our skin, lungs, kidneys, liver and brain. Cyanides cause rapid respiration, drop in blood pressure, dizziness, coma and are found in over two thousand plants, including almonds, cassava and sorghum. Furocoumarins cause issues with the skin and are found in citrus fruits and parsnips. Pyrrolizidine alkaloids cause damage to DNA and are found in 600 plants, including herbal teas, honey, herbs, spices, as well as grains.[119] Nightshades, which are types of alkaloids, cause issues in some people with digestion, skin, joints, muscles, respiration, etc. and are found in tomatoes and potatoes.[120] We evolved to taste toxins as bitter to avoid or minimize them in our diet.

Almost all recreational drugs are secondary plant compounds that affect the brain, including nicotine, caffeine, THC, CBD, psilocybin from mushrooms and mescaline from peyote. And most pharmaceutical drugs, both prescribed and not, are also secondary plant compounds. While these compounds are causing harm to us when we use them, they are nonetheless also providing other forms of benefits for us in certain conditions.[121]

Though not "secondary plant compounds," another set of compounds also interferes with the digestion of protein, called storage proteins.[122] While evolution designed these proteins to store amino acids, it also probably designed them to work as antinutrients by making them difficult to digest, as well as being toxic in some cases. Gluten falls into this category—but nearly all seeds contain some version of it to one degree or another. As noted before, fruits and berries contain fewer or no secondary compounds in their pulp; additionally, they make their pulp especially tasty with sugars and organic acids.[123] This is all to lure animals to eat their fruit and then deposit the seeds in their stool in some distant place—thus helping the plant to propagate.

ANCESTRAL CONSUMPTION

We have already seen that throughout evolution our ancestors strove to minimize the effects of these compounds. Many animals consume enormous amounts of secondary compounds every day — so they are constantly evolving and adapting to minimize their impact. Some animals, such as ruminants and even some primates, like the colobus monkey, contain forestomach's with microorganisms that ferment their food while also deactivating many of these compounds. But most primates have guts similar to ours, so they evolved other ways to deactivate these compounds. Some strive to avoid these compounds in the first place by focusing more on fruits and animals. Some eat many different kinds of plants, maybe twenty or forty during the day, so they avoid eating too many of any particular compound — which then helps them to avoid most detriments. Others eat certain plants only at certain times of the year when they are lower in these compounds.[124] Others evolved livers to deactivate the compounds in some of these foods, especially if they regularly consume them. Additionally, most primates consume soil and clay with their food, which binds with these compounds and prevents their uptake.

Our line of evolution, in particular, chose more refined foods in the first place, including fruits and animals, and later deactivated these compounds with techniques like fermenting and cooking. Agriculturalists bred their plants to contain fewer of these compounds — as all domesticated plants are less toxic than their wild versions.[125] Moderns continued the trend but in the process stripped away many nutrients, causing more harm than good. And with all these techniques, we perhaps overcame plant defenses better than any animal on earth.

ESTABLISHMENT RECOMMENDATIONS AND SCIENTIFIC STUDIES

While recognizing the existence of secondary compounds, the establishment does not recommend limiting them while, paradoxically, recommending diets high in whole plants, including whole grains, legumes, nuts and vegetables that are high in these compounds. So, if following that diet, many humans are likely over-consuming these compounds to their detriments.

Scientific studies obviously confirm the existence of these compounds, their effect upon our physiology, as well as ways to reduce them. Roasting, soaking, germinating and fermenting reduce these compounds,[126] with sprouting and fermenting seeming the most effective in reducing multiple compounds, including phytates, lectins, tannins, oxalates, saponins, goitrogens, alkaloids, glucosinolates, protease inhibitors, solanines, etc.

Interestingly and admittedly, some studies show possible benefits as well to some of these secondary compounds, like phytates and lectins,[127] but I surmise that those benefits are optimized only when we consume limited amounts of them—which will always happen as long as we are eating whole plant foods.

RECOMMENDATIONS

So this diet recommends you eat whole plants while deactivating their secondary compounds by following the patterns of our ancestors. First, you should select plants lower in these compounds, and second eat many different types of plants, including various grains, to avoid consuming too much of any one of these compounds. Third, you should properly process your seeds through soaking, sprouting or fermenting—and then thoroughly cook them.

You should remove the blemishes and sprouts from your potatoes as they are dense in nightshades and exercise caution about eating too many vegetables raw, especially bitter ones. While containing great benefits, kale for example contains at least eight secondary compounds that potentially damage our health, especially in higher amounts. You do not need to worry about fruits as they are low in these compounds even in the wild.

CHAPTER 20

PREMISE 4 – CONSUME ALL NUTRIENTS

The optimum diet recommends that you consume all nutrients to prevent deficiencies and to avoid making or converting them inside your body. This includes all macronutrients, vitamins, microminerals, as well as fiber and phytonutrients. When eating foods we are really eating nutrients—even though most of us do not think that way—because once we put food into our mouths, it starts to transform into nutrients that provide for our various cells and their metabolisms. So we of course should design our diets to balance our nutrients.

Despite this, we Americans seem fixated on treating food and nutrients in the same way as we treat politics, religions, sports and other endeavors: as divided between good and evil, democrat and republican, or my team versus that other team—or as good nutrients and bad nutrients, or good fats and bad fats, or good plant foods and bad animal foods. But in reality, all nutrients are good without exception as they all serve some vital function in our body. And as we well know, when not acquiring certain nutrients or enough of them, we merely make them inside our bodies. So this diet contains all nutrients and, as explained in the next section, all nutrients in their best ratios to each other.

All our ancestors, in general, strove to attain all nutrients in their diet, both the essentials and non-essentials, as long as they were available in their ecosystem. However, some organisms make most of their nutrients inside of themselves, like plants, which once they photosynthesize glucose, then convert all other nutrients from that one molecule. Some animals are carnivorous—so they only receive protein and fat in their diet and only trace amounts of carbs—so they make their own glucose and ketones. Ruminants convert polyunsaturated fatty acids into saturates and monounsaturates inside their gut; and convert many amino acids as well. However, nearly all of our ancestors are omnivorous primates that evolved to receive just about all of their nutrients from the diet, from mixtures of plants and animals—and only make nutrients when necessary, as during famines. However, as we know, there were some exceptions to this rule, namely carnivorous human foragers and herders.

While accentuating the importance of essential nutrients, the establishment nonetheless recommends that we eat "balanced" or "well-rounded" diets or all foods on the food pyramid, so they are essentially recommending that we eat essential and non-essential nutrients. They also recommend upper and lower limits for most nutrients—which keeps conversions lower. I have not seen any scientific studies comparing the diets of those who only eat essential nutrients and those who eat all nutrients. However, scientists study the effects of certain diets high in some macronutrients and lower in others, such as high protein or carb diets—but generally, as I will address later, these diets are not proven to provide any benefits and are more likely to cause detriments. This is because we evolved to acquire all nutrients in our diet and only make them when we could not find them, usually due to drought, fire, or famine. So in this case, all of our methodologies are in alignment or at least mostly so.

RECOMMENDATIONS

So this diet recommends that we consume all nutrients in our diet—which is easy because all these nutrients tend to clump into certain foods. For amino acids and fatty acids, we should consume fatty meats, seafood, eggs, dairy, grains combined with legumes or nuts and some form of collagen. Additionally, we should eat whole plants rich in carbs and fiber, as well as some fruits and vegetables that contain some nutrients not normally found in the rest of our diet. Generally, we should eat some additional forms of fat as well as salt.

CHAPTER 21

PREMISE 5 – OPTIMUM NUMBER OF CALORIES

This diet recommends that we consume the optimum number of calories. When eating the right number of calories, we tend to optimize our metabolism. When over-consuming calories, we tend to gain weight and trend ourselves towards the diseases of civilization. When under-consuming calories, we tend to slow down our metabolism but also, possibly, lose weight and delay aging through inducing autophagy. But overall, this diet recommends that you eat the right number of calories to optimize your metabolism—and I am guessing that this, too, also helps with weight loss and slowing aging.

NUTRITIONAL PHYSIOLOGY

Through the decades scientists have devised some effective, but not altogether perfect, methodologies for determining the number of calories in our diet. To understand this methodology, let's first define calories. By definition, calories are the amount of energy in food that then provides ATP for our body. To determine the number of calories in any gram of macronutrient, like glucose, scientists literally burn that macronutrient with fire and determine the amount of heat released and then correlate that with calories. Using this method scientists determine that protein yields four calories per gram (even though protein is not normally used to make ATP); carbs yield four calories; and fats yield nine calories per gram. Calories are an odd and analogous form of measurement because we humans do not actually burn calories in our bodies, but metabolize them, which are two different chemical processes—but evidently, the two are similar enough so that this measurement works well enough.

Scientists use three forms of measurement to determine the number of calories metabolized by our entire bodies over one day—or TEE (total energy expenditure). In direct calorimetry, scientists place people in sealed environments and then measure the heat produced by their bodies. Using various calculations, they then equate that heat to calories utilized by the body. We generate most of our heat from catabolism, but also from anabolism, as well as electricity and other mechanisms—so this heat only approximates catabolism. In indirect calorimetry, scientists isolate people and then measure their intake of oxygen and outtake of carbon dioxide, and from that data calculate their calories utilized. And in this case scientists only measure oxygen from one form of catabolism, the Krebs cycle. Using this same measurement, scientists also determine the Respiratory Quotient—that is, the relationship of oxygen uptake to carbon dioxide outtake—which then determines the ratio of protein, carbs and fats that are catabolized in our body. Finally, in doubly labeled water, scientists label water with isotopes—and then measure the excretion of that water, along with carbon dioxide, and from this data determine the number of calories. Obviously, none of these measurements is perfect but only suggestive.

Scientists then further break down these whole-body metabolisms into the different categories already mentioned in brief: basal or resting metabolism; digestive metabolism (the thermic effect of food); and exercise metabolism. Scientists can measure these specific metabolisms either with direct or indirect calorimetry as discussed above. To determine basal metabolism, people have their heat and gases measured at rest; to determine digestive metabolism, people are measured while eating food, and then scientists measure the increase in their heat and gases over basal—and the same process for exercise metabolism.

We use about 60 to 80% of TEE for basal metabolism with our organs requiring about 60% of that amount. And as mentioned, some organs are expensive, using way more metabolism than others, such as our brain, liver and kidneys. Meanwhile, our muscles require 20% of our basal (mostly just to generate heat) and our body fat requires about 10%.

Generally, we humans have somewhat similar basal metabolisms for two reasons: we all need to maintain the same amount of heat to keep us alive, and our organs all need to perform the same functions at rest. But we also vary considerably due to our differences in lean mass—the more mass, the more metabolism. For this reason, women and the elderly tend to have lower basal metabolism.[128,129] But paradoxically, as discussed in prior sections, they also have higher basal metabolism per cell due to thermogenesis—that is, they have more surface area relative to mass, so they bleed more heat and thus need to generate more heat by raising their basal.

We also vary our basal for other reasons—but only slightly. First we vary because of our environment: the colder our environment, the more heat we generate inside our bodies to compensate; and the reverse—even though we can always just wear more clothes. Some evidence suggests that some animals, including humans, evolved variations in basal to adapt to temperatures in their native environments; arctic people thus have higher basal metabolisms and tropical people lower basal metabolisms—even as these variations seem minimal.[130] Accordingly, humans and other animals in colder environments tend towards greater mass and thickness to adapt them to the cold—and the reverse for the tropics—although other factors also determine the shape of our bodies. Second, our basal varies depending on our percentage of body fat: with greater body fat, we lower our basal because fat cells metabolize fewer calories as compared to muscle cells—about one third less.[131] We also insulate better and thus conserve more heat. Third, and as mentioned before, our basal varies on our intake of calories: if we reduce our calories, we eventually lower our basal by as much as 10%;[132,133] if we raise our calories, we raise our basal. Finally, our basal varies around exercise;[134] when exercising more, we produce more basal because after the exercise our metabolism stays elevated and we need more metabolism to repair our tissues.[135]

One study showed that amongst 150 adults, basal metabolism varied from about 1000 to 2500 calories per day, with the average at 1500. About sixty percent of the differences were due to lean mass — or in other words, just the size of people. Seven percent was due to the percentage of fat mass. And the remaining 20% was not due to differences in the size of the brain or organs, but to other causes, presumably some combination of those listed above.[136] So, in short, once adjusting for lean and fat mass, we humans vary in our basal somewhere around 20% — although most of us, probably less than that. Oddly enough, basal metabolism of humans has fallen about 6% over the past thirty-five years.[137]

Digestive and active metabolism

Digestive metabolism requires about 10 to 30% of your whole-body metabolism — which I already addressed. Active metabolism varies radically, dependent upon the amount of exercise. On the lower end, humans can have next to zero active metabolism if they are, for example, sofa surfers. On the upper end, however, humans have active metabolisms as high as 2.5 times their basal metabolism but not any higher: usually this is professional endurance athletes.[138] So, if your basal metabolism requires about 600 calories per day, you can only use about 1500 calories for active metabolism. But under most and average scenarios, we humans distribute our whole-body metabolism along these lines: 60 to 75% basal, 10% digestive, and 15 to 30% active. However, some recent studies are questioning the way active metabolism works in relation to basal metabolism.[139]

ANCESTRAL CONSUMPTION

While evolving our ancestors increased the mass of their bodies and brains, so their overall basal increased, even while decreasing some for thermoregulation. In colder environments they had higher basal rates; and lower in the tropics. They also decreased their digestive metabolism due to consuming more refined, processed and predigested foods. Their exercise metabolism probably remained somewhat the same all the way through agricultural humans. As moderns, however, we lowered our basal overall for various reasons; increased our digestive metabolism because we overate even while eating more refined foods, and then sharply reduced our exercise metabolism.

ESTABLISHMENT RECOMMENDATIONS AND SCIENTIFIC STUDIES

From all their studies on metabolism, using direct and indirect calorimetry, as well as "doubly labeled water," scientists have observed the metabolism of humans over decades. And from these observations, they tabulated and analyzed enormous amounts of data to draw conclusions about our TEE. From these studies they then approximated the amounts of calories people need per day and then compared those amounts to the amounts consumed by people in reality, finding they generally aligned. They then used that data to generate "calculators" that help anyone determine within seconds and with some degree of accuracy the number of calories needed for TEE (total energy expenditure). There are several of these calculators based on calorimetry with slight differences in their methodologies: The Schofield equation,[140] The Harris-Benedict Equation,[141] and The Revised Benedict Equation, but The Mifflin-St. Jeor Equation[142] is considered most accurate even as they all tend to deliver similar results. However, the Institute of Medicine Equation[143] is based on doubly labeled water and is considered the most accurate.

These calculators work as you might expect. First, they try to determine your basal metabolism by asking for your weight and height to determine your mass-to-fat ratio. For more precise results, some ask for your actual body fat (like the Katch-McArdle calculator[144]—which you need to measure separately through several methods. Additionally, these calculators ask for your sex and age even though these do not influence your basal all that much, if at all. And with this information, these calculators approximate your basal metabolism with some degree of accuracy over one day.

They then calculate your active metabolism by asking for your level of activity—that is, the time and intensity of exercise, and from that, they calculate your exercise metabolism.

They then automatically calculate your digestive metabolism based on data I presented earlier. Given all the variations in people and metabolisms, and the shortcomings in these methods, these calculators are probably accurate at somewhere around 10 to 30%, depending on various factors.

To determine calories in this book, I use the concept of the theoretical consumer. This is essentially you—the reader of this book. But we all come in different shapes and sizes, so for that reason, I decided to average us all out into the theoretical consumer that is about thirty-five years old, neither male or female, while weighing about 150 pounds and standing about five feet and ten inches while exercising moderately about five times per week. In some cases, I also estimate the person has about 15% body fat—and thus is reasonably fit and presumably active. I then plug this consumer into our calculators to determine TEE in calories. And from there, I can determine the amount of protein, fats and calories this person needs, in g, and as a percentage of total calories. The nutritional establishment uses this same theoretical consumer to make their own nutritional analysis for the labels on the backs of packaged foods.

In any case, when inputting the data for our theoretical consumer into these various calculators, I receive the following results for basal alone.

- The Harris-Benedict equation: 1621 cal/day

- The Revised Benedict equation: 1628 cal/day

- The Mifflin-St. Jeor equation: 1597 cal/day

- The Schofield equation: 1655 cal/day

- The Katch-McArdle equation: 1600 cal/day

When adding digestive and exercise metabolism to this basal metabolism, our theoretical consumer needs somewhere around 2300 cal/day. However, the Institute of Medicine Equation, the one based on doubly labeled water, gave the result as 2800 cal/day. To make your own calculations, you of course need to determine your weight, your body fat and your level of exercise and then input that data. Additionally, for better results, you can determine your body fat and use that data as well.

RECOMENDATIONS

As noted before, the purpose of this diet is to help you optimize your metabolism. To accomplish that goal, you should first estimate the number of calories you need per day, or TEE, by using one of the calculators. You then need to determine if you are eating those calories—which means you need to count the calories in your food by using labels and research. You can then compare the theoretical to the actual calories and note the difference, all the while acknowledging that this process is not perfect, but approximate, given all the variables involved.

If your actual is substantially over your theoretical, you are probably eating more calories than you need, especially if you are overweight. You then want to consider if you are overeating because you consume too many over-processed foods, especially those rich in carbs and fats. Or maybe you are not eating enough protein, causing you to overeat carbs and fats, as explained before? Or maybe you are not exercising enough? Otherwise, you are possibly overeating for many reasons beyond the scope of this book, including hormonal imbalance, entrenched habit or psychological distress. And when asking these questions, keep in mind we do not have much, if any, reason to believe that humans evolved to eat themselves into obesity.

If your actual is substantially under your theoretical, you are possibly not eating enough calories to optimize your metabolism—and thus living in compromise with lower catabolism and anabolism. So you need to evaluate the reasons for this: maybe you are afraid of gaining weight, and thus constantly restricting yourself unnecessarily; maybe you are concerned about the costs of your food or you are eating the wrong foods that then dampen your appetite without you knowing this? Maybe you are consuming too many secondary plant compounds? Too much fiber? Or perhaps your fatty acids and amino acids are imbalanced?

If your intention is not to optimize your metabolism, but to lose weight, you want your actual lower than your theoretical—which then triggers you to pull more from your own fat. Or in other words, you want to consume fewer calories, but based on nutritional physiology, we do this best by eating as much protein as we need; by eating as many whole carbs as we need but not any more; and by restricting fat—while also exercising at low to moderate levels where your muscles catabolize fats.

If your intention is not to optimize your metabolism, but to slow aging through autophagy and other triggers, you again want your actual lower than your theoretical calories. In this particular case, you probably want to lower your calories across all macronutrients because you are trying to slow down your basal metabolism while also inducing autophagy. Because you are eating less, you lower your digestive metabolism, especially if your foods are processed properly. And you should intentionally reduce your exercise metabolism. Theoretically this will extend your life but likely compromise your life as well because you are living lower on both catabolism and anabolism. To avoid this condition, you can probably best extend your life while maintaining your metabolism by eating the diet presented in this book.

CHAPTER 22

PREMISE 6.1 – OPTIMUM AMOUNT AND RATIOS OF NUTRIENTS: PROTEIN

This diet also recommends that we consume all nutrients in their proper ratios to each other—that is, the way our body uses nutrients for our various metabolisms. All nutrients are neither inherently good or bad; however, they become bad or good depending on the amount consumed and their ratio to other nutrients. When over-consuming certain nutrients, we usually cannot utilize them and for that reason we either convert or store them. When under-consuming nutrients, we still need those nutrients, so we then must convert them from other nutrients. But when consuming nutrients in their proper amounts and ratios, however, we consume, digest, transport, uptake, store and metabolize only the nutrients we need—and do not waste all that on the nutrients we do not need. So the question is: what are these proper amounts and ratios of our macronutrients? In the pages ahead, I will answer that question, starting with protein.

NUTRITIONAL PHYSIOLOGY

On average we humans utilize about 300 g of amino acids per day for anabolism, about the equivalent of sixty eggs, but most of that comes from recycled protein. When proteins are damaged in our body, our enzymes disassemble them and then discard the damaged amino acids and recycle the healthy ones. So for anabolism we receive around 80% of our amino acids from recycled protein and only 20% from dietary protein.

When consuming too little protein, we slow down our anabolism and otherwise utilize our own muscles for amino acids.[145] In this case we also tend to over-consume other macronutrients, namely carbs and fats, to try to attain enough protein. When consuming too much protein, we tend to oxidize or convert the amino acids into glucose or ketones while wasting our metabolism on processes we do not need. We specifically waste digestive metabolism, given that protein requires so much metabolism to digest. We also release more wastes such as urea, ammonia and uric acid, which are toxic at higher levels.[146] However, when consuming the proper amounts of amino acids, we provide our cells with what they need for proper anabolism — without any downside.

Generally, all humans need similar levels of protein per cell for basal metabolism. However, obviously, if someone is bigger, containing more lean mass, they need more protein because they have more cells — even as, due to thermoregulation, they need slightly less protein per cell. We also increase our need for protein when increasing our level of exercise metabolism, as exercise tends to degrade our muscles, thus requiring more protein to rebuild.

ANCESTRAL CONSUMPTION

While evolving, our ancestors increased in size, and thus needed more protein but, again, they needed less protein per cell. They also evolved to consume more of their protein from animals relative to plants, and then process that protein—thus increasing the quality of both their complete and collagen proteins.

Although our predecessors probably consumed about the same amount of protein per mass, humans varied from around ten to 30% of calories for various reasons. First, they varied because they carried different amounts of muscle mass. Second, they varied because of the different amounts and degrees of exercise metabolism. Third, some had higher basal metabolism when living in the cold. But they varied most of all based on the availability of other macronutrients in their ecosystem—so that when their environments were low in carbs, they consumed more protein to compensate both for the reduction in calories and to convert that protein into glucose and ketones for their brain.[147]

As moderns we continued to eat about the same amount of protein, but more than needed given that we are so sedentary; across continents and time, most moderns consume about 16% of their calories from protein.[148] However, we lowered the quality of our protein so that we now over-consume muscle protein, under-consume collagen protein, and do not balance the various seeds in our diet to generate complete proteins.

ESTABLISHMENT RECOMMENDATIONS AND SCIENTIFIC STUDIES

The establishment recommends that humans consume about 0.8 to 2.0 g per kilogram of lean mass[149]—which amounts to somewhere between 8 to 30% of our total calories from protein—an enormous range, from the lowest to the highest possible. They base their recommendations for protein upon the methodology known as "nitrogen balance." Since protein is the only macronutrient that contains the atom nitrogen, scientists can track protein by tracking nitrogen. In simplified terms, they track nitrogen intake in our diet and nitrogen excretion in our urine, feces and sweat—and from that data, conclude how much protein we need per day—even while this methodology is imperfect for various reasons.

In positive nitrogen balance, humans are eating more protein than they need just for maintenance—and are thus growing their tissues, as is the case with children and athletes. In negative nitrogen balance, nitrogen intake is less than nitrogen excretion. In this case, humans are not consuming enough protein—and thus are breaking down their muscles. In nitrogen balance, nitrogen intake from our food equals nitrogen excretion. In this case we consume enough protein to maintain our tissues in balance. Obviously, most humans want to strive for nitrogen balance.[150]

Through studying nitrogen balance, scientists have thus determined, with some degree of accuracy, the amount of protein that humans need on average per day, for basal metabolism—which is then adjusted for other variables like lean mass and exercise metabolism. They then cross-reference these amounts with observational studies—that is, basically measuring the amount of protein people consume in real life. And in the end, these two measurements are generally congruent with each other.

From all this data scientists generally conclude that humans achieve nitrogen balance somewhere between 0.6 g to 0.9 g of protein per kilogram of weight.[151] At the low end of this range, some humans are consuming enough protein for nitrogen balance and thus avoid breaking down their own muscles. But they are possibly slowing down their anabolism, which then probably slows down catabolism as well.

When consuming more protein than this, humans receive considerable benefits. With more protein, we feel more sated and thus eat fewer carbs and fats. We also produce more of the hormone, glucagon, as compared to insulin, which in turn helps us better utilize our body fat and prevent weight gain; this is the premise behind most of the high-protein diets like the Atkins and Zone and many others. Additionally, humans tend to improve cardiovascular function, wound healing, bone health, muscle mass, strength and function, and maintenance of energy balance. Humans also reduce obesity, along with type 2 diabetes, metabolic syndrome, heart disease and sarcopenia. They lower their triglycerides and total cholesterol along with LDL—while increasing their HDL as well as their immunity. And with more of the amino acid, leucine in particular, humans appear to function better with cell signaling, satiety, thermogenesis, and glycemic control.[152]

Additionally, more protein does not cause loss of calcium through the urine but helps retain calcium and enhance the turnover of bone. In general, many studies have suggested benefits to consuming protein over and above the lower amounts recommended by the establishment.[153]

However, at amounts somewhere around 1.2 g per kilogram or so, most people are entering positive nitrogen balance,[154] eating too much protein. At this point, they still receive the benefits mentioned above but do not enhance them and perhaps start to negate them and otherwise cause the detriments associated with the over-consumption of protein.[155] Then, at somewhere around 25% or so of calories, humans start trending towards pathology, essentially consuming more amino acids than the body can tolerate, resulting in enormous amounts of conversion and wastes, while trending towards hyperammonemia, osteoporosis, hyperaminoacidemia, hyperinsulinemia, dehydration, nausea, seizures, elevated homocysteine levels (i.e. CVD risk-factor), increased calcium excretion, issues with the kidneys and liver and even death.[156] For this reason, our ancestors never consumed protein over 30% of total calories, which is equated to about 2.5 g of protein per kilogram of weight.

So even though, the nutritional establishment recommends that protein can range between 0.6 to 2.0 g per kilogram of weight, and that we can achieve balance somewhere around 0.8, we are probably best served with protein somewhere around 1.0 to 1.6 g of protein, depending on our lean mass compared to fat mass, as well as the intensity and length of our exercise. When calculating these numbers for our theoretical consumer, we get the following results while also showing the extremes:

Theoretical Consumer: 150 pounds, active and fit
Calories: 2300 per day
Protein per day in g and percentage of total calories:
- 0.6 g equals 41 g of protein (7% of calories)

- 0.8 g equals 55 g of protein per day (8%)

- 1.0 g equals 68 g of protein (10%)

- 1.2 g equals 81 g of protein (12%)

- 1.3 g equals 88 g of protein (13%)

- 1.4 g equals 95 g of protein (15%)

- 1.5 g equals 102 g of protein (16 %) (about

- 17 eggs per day)

- 1.6 g equals 108 g of protein (17%)

- 1.8 g equals 122 g of protein (19%)

- 2.0 g equals 136 g of protein (21%)

RECOMENDATIONS

When consuming protein at under 8% of calories, we humans are possibly pulling amino acids from our muscles. At somewhere around 8 to 10%, we are probably maintaining our mass but not optimizing our anabolism. At somewhere around 10 to 16%, we are probably receiving optimum amounts of protein. At over 16 to 25%, we start to convert protein into other nutrients. At 25 to over 30%, we experience detriments—essentially protein toxicity—otherwise known as rabbit sickness, the term coined by frontiersmen. So based on establishment recommendations, further assessed through scientific studies, you should consume protein somewhere between 1.0 to 1.6 g per kilogram or somewhere between 10 to 16% of calories from protein. At that point you are replacing discarded amino acids with dietary amino acids in their proper ratios.

Generally, most of our ancestors consumed on the higher end of this amount, or even somewhat higher, because of their higher exercise metabolism and other factors. Some consumed more than that, but only in ecosystems low in carbs. So this range accords with nutritional physiology, ancestral consumption, as well as establishment recommendations and scientific studies—even as some pop gurus suggest we need more protein. And as noted, nearly all moderns consume about 16% of their calories from protein, on the high side of optimum.[157]

But how do we moderns further define this range for ourselves? You should consume on the lower end of this range if you are lower in lean mass and exercise metabolism—and the reverse. As noted before, with greater exercise metabolism, you radically increase your need for calories from carbs and fats but only marginally increase your need for protein. Endurance athletes only need between 1.0 to 1.6 g of protein per day; if you are in the low to moderate zone in your exercise, you need on the lower side of this range; if you are high zone, say competing in the Tour de France, you need as much as 1.6 g of protein.[158] Interestingly, as an overall percentage of total calories, protein actually decreases substantially with exercise—because you are consuming lesser amounts of protein relative to carbs and fats.

Some athletes need more protein because they are breaking down more tissues in sports like weightlifting, football, rugby, and tennis—somewhere in the range of 1.6 to 1.7 g per kilogram on the upside with no benefits resulting from more than that.[159] If you restrict carbs and fat and eat more protein, however, you are more likely to gain more lean mass, even though that mass does not help your performance.[160] Interestingly, eating protein immediately after exercise does not appear to enhance recovery.[161] In extreme cases, like bodybuilding, some may benefit from over 16 to 20% of protein. But with all of this said, you need less protein than mentioned above overall when you consume complete proteins, and then even less when you combine those completes with collagen. So in that case you might just need more like 0.8 to 1.0 g of protein per kilogram per day.

If you want to lose weight, you should probably eat in the middle to high end of this range of protein for several reasons. First, you cannot gain weight from protein because for that to happen, protein must first convert into glucose and then into palmitate. Second, you gain more muscle from eating protein, which in turn helps you burn more calories. Third, you burn more fat from eating protein—thus helping to lose weight in the short term, but not in the long term because 90% of people gain that weight back.[162]

CHAPTER 23

PROPER RATIOS OF AMINO ACIDS

NUTRITIONAL PHYSIOLOGY

This diet also recommends that we not only eat protein but that we eat our amino acids in certain ratios to each other. When this happens, we then replace discarded amino acids with dietary amino acids in the same approximate ratios to each other. To understand this point metaphorically, let's think of proteins as puzzles that consist of twenty-one pieces of amino acids. When one or more of those pieces is missing, we cannot complete the puzzle; we then need to make those pieces or pull them from another puzzle. When containing more pieces than needed, we then cannot use those pieces and in some senses waste them or turn them into other pieces that do not even fit our puzzle.

When consuming amino acids in improper ratios, or, more specifically, when consuming too few of one or several amino acids, "the limiting amino acids," we cannot anabolize our proteins.[163] When this happens, we compensate in two ways: we make those limiting amino acids inside our bodies from precursors—but with the usual detriments already discussed. But in many cases, those amino acids are essential so we cannot make them at all—and in this case, we then must pull them from our muscles.

When consuming too many of certain amino acids, we cannot utilize those amino acids so we must rid ourselves of them in the usual ways. But when consuming the proper amounts of amino acids, we then anabolize them in their correct ratios without any downsides. So we want to replace discarded amino acids in the same ratios with our dietary amino acids—at least approximately.

How do we do that? As already noted, we humans make many compounds from proteins in our diet, including peptides, enzymes, and proteins. While there are thousands of proteins in our body, most of them fall into two categories that contain similar ratios of amino acids—that is, complete proteins, found in muscle and most of our organs, that account for about 70% of our protein. And collagen proteins, along with keratin and elastin proteins, that account for about 30% of our protein, located in such tissues as skin, hair, nails, ligaments, tendons, fascia and embedded throughout other tissues like our nervous, pulmonary and cardiovascular systems. So we humans are somewhere around 70% complete proteins and about 30% collagen proteins.

As compared to complete protein, these collagen proteins are dramatically higher in the three amino acids, glycine, proline and hydroxyproline, but dramatically lower in the nine essential amino acids found in complete proteins. Or in short, these two types of protein contain very different ratios of amino acids—and thus, to convert complete into collagen proteins requires enormous amounts of conversions. For this reason, we want to consume these classes of proteins in the same amounts as used in our body—that is, at 70% completes and somewhere around 30% collagen. (However, this does not, and can not, account for differing rates of turnover for these proteins because scientists do not know these rates.)

ANCESTRAL CONSUMPTION

Most carnivorous and omnivorous animals consume all of their prey—the organs, the fat, the muscles and all of the connective tissue, including the skin, fur, ligaments, tendons and in some cases even the bones. And in the process, they consume complete and collagen proteins in percentages that approximately match their own. Our common ancestor consumed whole insects, invertebrates, shellfish and occasionally the entirety of larger animals, and this behavior probably continued through Australopithecus and human foragers who consumed all of the animal, including parts of larger animals, like the hoofs, tongues, tails, skin, etc.[164] In some cases, however, they did not eat all of the animal and gave the muscle meat to their dogs or turned the skin into hides.

Agriculturalists consumed collagen mostly as broth, sausages, skins and pickled snouts and feet and the like. In many cultures broth was the basis of most meals as soups, stews and sauces. Moderns slowly reversed this trend, however, and focused only on complete proteins found in muscle proteins, like steaks, chicken and duck breasts, as well as in eggs and dairy, because the nutritional establishment advised against them, which I will discuss shortly, and we started to find them unappealing. Most of us only consume about 3 to 23 g of collagen per day.[165] Additionally some continue to consume Jello which is made from gelatin (one form of collagen), and many now are supplementing with collagen peptides and similar products. But modern Americans on average under-consume collagen, especially as compared to completes.

ESTABLISHMENT RECOMMENDATIONS

The nutritional establishment mostly uses the methodology of nitrogen balance as well to determine the quality of the ratio of amino acids in various forms of protein, like eggs, meat or beans—which they then rank as either lower in quality or completeness or higher in quality or completeness. When scoring higher in quality, food proteins contain ratios of amino acids that are better utilized by the body—or to use my own terminology, that better optimize anabolism.

With Protein Efficiency Ratio (PER), scientists feed rats specific kinds of food proteins in the same amounts and see which ones cause the rats to grow the most.[166] Obviously, when causing more growth, that particular protein is ranked higher. With Net Protein Utilization, scientists fed humans various food proteins and identified which ones produce the most amount of nitrogen balance—which signified that more of the amino acids are utilized and fewer are converted.

With Biological Value, scientists rank food proteins on the ratios of only their essential amino acids—while ignoring the non-essentials on the assumption that our body can make them. Many decades ago, scientists isolated the nine essential amino acids, mixed them in various ratios to each other, and then fed them to children and animals and determined which ratios generated the most nitrogen balance—and then defined those ratios as the standard. Oddly enough, in their next step, they then compared this standard to only the essentials in certain foods like eggs, meat and beans.[167] The closer those ratios matched, the higher the quality of the protein; the less they matched, the lower the quality of the protein.

While regarded as the best of the standards, Biological Value is flawed because it focuses only on the essential amino acids while ignoring the non-essentials altogether, assuming that converting amino acids is just as good as receiving them in your diet. However, this is not the case for several reasons. First, we are limited in our ability to make the non-essential amino acids, especially if the precursors are low, or we are under stress.[168] We can only convert 30% of the non-essential glycine (one of the predominant amino acids in collagen) that we need. While needing about 10 g of this amino acid every day, we can only make about 3 g while only receiving about 3 g per day in our diet from complete proteins, leaving deficits of somewhere around four to 6 g per day. In other words, if we do not consume enough glycine, we do not meet our needs, resulting in this case in a reduction in collagen in the body.[169]

Second, consuming both essentials and non-essentials is better for anabolism and in rats in particular, the best ratio of total amino acids to essential amino acids was three to four.[170] Third, in PER studies, animals absorb more nitrogen/amino acids into their tissues for growth in the presence of both essential and non-essential acids.[171] Fourth, we do not eat any proteins with just the essential amino acids because they do not exist in nature, only in labs. Finally, Biological Value is flawed because it does not corroborate its own conclusions about protein quality with any other biomarkers.

Despite the various strengths and weaknesses of all these methodologies, they all tend to rank proteins about the same on their quality or completeness, with eggs and dairy at the top of their list; this makes sense given that these foods evolved to optimize the anabolism of growing animals. These are followed by soy and various muscle meats like chicken breasts or steaks. Nearly all plant proteins, with the exception of soy, score low in these rankings because they contain limiting amino acids—that is, the one or two amino acids that are low relative to the other amino acids, and that prevent the making of the complete protein. For most grains the limiting amino acid is lysine—which means that grains alone are poor sources of protein. However, lysine is high in most legumes, but their limiting amino acid is cysteine which is high in grains. So when combining grains and legumes together, they compensate for their limiting amino acids and dramatically raise the overall quality of the protein to the top of the list, assuming those proteins are absorbed by the body.[172]

Here is one ranking of protein quality and completeness:[173]

- Whey Protein: 96

- Whole Soy Bean: 96

- Human milk: 95

- Chicken egg: 94

- Soybean milk: 91

- Buckwheat: 90

- Cow milk: 90

- Cheese: 84

- Quinoa: 83

- Rice: 83

- Defatted soy flour: 81

- Fish: 76

- Beef: 74.3

- Immature bean: 65

- Full-fat soy flour: 64

- Soybean curd (tofu): 64

- Whole wheat: 64

- White flour: 41

- Gelatin: 30

When combined, plant proteins score much higher by compensating for their limiting amino acids. However, these scores do not account for the lower digestibility of plant proteins:[174]

- 85% rice and 15% yeast: 118

- 55% soy and 45% rice: 111

- 55% potatoes and 45% soy: 103

- 52% beans and 48% corn: 101

As you can see, the nutritional establishment scores collagen protein at the bottom of the list. So while recommending that we consume "complete" or "quality" proteins, they also recommend that we avoid collagen proteins altogether or at least regard them as low in quality.

DIGESTIBILITY OF PROTEIN

For the digestibility of proteins, scientists rank food proteins on the completeness of their digestion, using nitrogen again as their measurement. In this case, they control the diet of humans, count the nitrogen in their diet and then count the nitrogen in their stool or at the end of their small intestine and then subtract the two from each other, with the difference being the amount of nitrogen (or protein) that was not digested.

They then rank the proteins based on their digestibility. Animal proteins are more digestible in almost any form, and plant proteins are less digestible because of their fiber, antinutrients, toxins and storage proteins.[175] They are also possibly less digestible due to their lower quality proteins, as we may not absorb amino acids we cannot utilize well. But other factors also contribute to digestibility, including, most of all, how we predigest both plant and animal protein. If seeds are sprouted or fermented and ground, and then well-cooked, they possibly then become as digestible as animal proteins.[176] All animal proteins digest well even when raw, but they digest more efficiently when cooked and ground, or otherwise, cured, fermented and aged. In some cases, too, food proteins are more digestible when combined together, as with meat and skin or, said another way, complete and collagen proteins. Interestingly, gelatin is ranked as high as most other animal foods in digestability.[177]

Here are the digestibility rates of several foods:[178]

- Animal proteins: 90 to 99%

- Various Mixtures of beef and pig skin: 95 to 100%

- Gelatin (collagen) 90 to 95%

- Cheese: 95%

- Eggs: 97%

- Tofu: 90%

- Plants: 70 to 90%

PDCASS AND DIASS

As noted before, the nutritional establishment uses PDCASS (Protein Digestibility Corrected Amino Acid Score) as their ultimate methodology for ranking food proteins in their quality. In PDCASS, scientists essentially combine Biological Value (the optimum ratio of essential amino acids) and protein digestibility—and then rank the food proteins on their quality.[179] Another methodology, DIAAS (Digestible Indispensable Amino Acid Score) uses the same methodology except it measures protein digestion at the small intestine instead of the large intestine.[180] Both generate about the same results—and in the end, both then rank proteins in about the same order as the other rankings so that all these rankings generate similar results.

SUMMARY AND RECOMMENDATIONS

Overall, these methodologies help to determine the quality of protein as well as its digestibility. But they only focus on complete proteins—generally the ratio of amino acids found in our muscles and organs—while ignoring collagen altogether. Furthermore, Biological Value ranks collagen at the bottom of their list because it's low in the essential amino acids and even devoid of some of them. But collagen is also high in the non-essential amino acids, including glycine, proline and hydroxyproline, the three that are low in complete proteins. So Biological value ranks collagen at the bottom of the list but this only means that collagen is low in quality in their flawed methodology, not in physiological reality. Overall, while nutritional physiology and ancestral consumption support the consumption of collagen, establishment recommendations does not. To attempt to reconcile this discrepancy, I consider scientific studies on collagen in the pages ahead.

COLLAGEN SAFETY

Since animals, including humans, have consumed collagenous tissues for millennia, we can assume that these foods are safe to consume as well as digested and assimilated—as animals do not evolve to consume foods that are bad or wasteful for them. Furthermore, scientific studies, even those performed by the FDA, confirm that collagen is safe without any detriments; gelatin, in particular, is safe for humans at doses of 10 g/day for three to six months.[181,182] They have also proven that enormous doses are safe for animals—but have not conducted those studies on humans.[183]

COLLAGEN DIGESTION

We appear to digest collagen as well as any other animal food. Other animals digest all forms of collagen at well over 90%, including skin, tendon, gelatin and peptides.[184] Humans do not digest raw collagen well at all—but when joints, skin, etc, are cooked in one form or another above 180 degrees, we digest them as well as any protein. Numerous studies show that humans digest gelatin and peptides and absorb them into their blood and then into various tissues—presumably at high rates.[185,186,187] Humans also absorb collagen protein as well as muscle proteins when they are mixed into fermented sausage.[188] Gelatin improves the digestion of milk,[189] beans, and meat while also increasing the utilization of the proteins found in meat and grains, including wheat, oats, and barley.[190]

EFFICACY OF COLLAGEN

When ingesting collagen in various forms, both animals and humans improve their skin. In rats collagen stimulates the anabolism of skin,[191] helps their skin retain water,[192] and improves their extracellular matrix.[193] Collagen reduces sun damage in hairless mice,[194] and enhances wound healing in diabetic rats.[195] In mice collagen stimulates the growth of fibroblasts in the skin and increases the number of fibroblasts migrating from the skin.[196] In guinea pigs Collagen increases the production of hyaluronic acid which is essential to skin and joints.[197] In humans collagen peptides stimulate the proliferation of both cells and hyaluronic acid in the fibroblasts of the skin.[198] These results suggest that Pro-Hyp might stimulate the growth of fibroblasts in the skin and consequently increase the number of fibroblasts migrating from the skin, improve dry skin and pruritus,[199] and improve skin elasticity.[200] In women, over six months collagen diminished the appearance of cellulite and improved skin elasticity,[201] and reduced wrinkles and roughness.[202]

BONES AND JOINTS

In mice collagen enhances the anabolism of bones.[203] In humans gelatin augments the synthesis of collagen throughout the body, especially in cartilage and joints and the knee in particular.[204,205,206] In rats gelatin improves calcium absorption and helps preserve bone strength and mineral density.[207] In human athletes gelatin reduced deterioration and pain in joints,[208] and improved joint pain and function in people with arthritis or similar conditions.[209] Peptides increased activity in humans with pain in their knees[210,211,212] and reduced and enhanced function in humans with problems in their Achilles tendons,[213] as well as in their ankles.[214]

TRAINING

When consuming gelatin three times per week for twelve weeks, resistance trainers showed increased muscle mass, the same amount of fat mass, and slightly greater strength. When consuming the placebo, they did not show any of these benefits and gained weight.[215] Elderly men and premenopausal women, with deteriorating muscles, showed improvements in strength and tone when supplementing with peptides and resistance training.[216,217]

CARDIOVASCULAR

Scientists have shown, at least in animals, that collagen protects against various forms of heart disease and cardiovascular damage.[218] In rats with high blood pressure, gelatin from chicken legs reduced their blood pressure and protected them from hypertension and cardiovascular damage.[219] In mice collagen reduced the size of lipoproteins, and in rats expedited the absorption and metabolism of lipids.[220] Finally, in mice gelatin improved glucose tolerance.[221]

BRAIN

In mice gelatin improved the functioning of the brain, including cognitive and spatial abilities.[222] In aged mice gelatin promotes the regeneration of the damaged brain and promotes learning and memory.[223,224] In humans collagen peptides significantly improved brain structure and cognitive function.[225] In humans gelatin promotes the secretion of GLP-1 that provides benefits to the brain.[226] Collagen also spans much of the brain and nervous system and helps prevent disease and promote homeostasis.[227]

MISCELLANEOUS

In animals gelatin increases red blood cells, hemoglobin, calcium in the blood, utilization of calcium and prevents the wasting of muscles.[228] In humans the production of collagen helps repair and generate tissues in the gut.[229] Collagen works as an antioxidant, neutralizing free radicals and also reducing inflammation.[230]

PERCENTAGE OF COLLAGEN TO COMPLETE PROTEINS

In nutritional physiology, I estimated that humans need somewhere around 10 to 30% collagen to 70 to 90% completes in their diet, based on several lines of reasoning. First, we are about 70% completes and 30% collagen in our own bodies—even as scientists do not know our rate of turnover. Second, we evolved to eat other animals in their entirety, which presumably contains these approximate ratios in their own body. I am guessing, too, that ancestral meals conformed to these parameters as well with such dishes as stews, soups, sauces, etc.

Additionally, scientific studies make suggestions about these ratios. When rats are fed muscle proteins with 15 to 22% collagenous proteins, the whole meal scores just as high in "biological value" as when they are fed muscle protein alone.[231] One study on humans affirmed the same. Using computer simulations of Biological Value, scientists showed that when increasing collagen from around 3% to around 18% in muscle meats, humans then better utilized all of the protein, suggesting that humans should consume about 80% complete to around 20% collagen.[232]

Meanwhile in another study, other scientists, based on calculations alone from PDCASS, showed that when consuming 30% of their protein from collagen and the remainder from complete proteins, humans do not experience insufficiencies in their consumption of essential amino acids. This further suggests that we humans should consume completes and collagens in the same percentages as found in our body.[233] In another study, older females who consumed as much as 50% of their total protein from collagen preserved lean mass and maintained nitrogen balance, while those consuming whey protein alone lost body mass and increased nitrogen excretion.[234] Another study showed that humans fed whey excreted more nitrogen than those fed collagen — even though both maintained nitrogen balance.[235]

RECOMMENDATIONS

Generally, all four of my methodologies recommend that we consume around 1.0 to 1.6 g of protein per kilogram of weight — or about 10 to 16% of overall calories.

If sedentary, with lower muscle mass, you should consume on the lower end of this scale. If more athletic, with greater muscle mass, you should consume on the higher end of this scale. Even if weight-lifting, you probably do not need more protein than the upper end of this scale. Interestingly, most moderns consume at the higher end of this scale, at 1.6 g — which is more than they need, given their sedentary lifestyle, but moderns over-consume all macronutrients to their detriment.

All four of my methodologies support consuming higher-quality or complete proteins. However, only three of my four methodologies recommend that we consume collagen protein, with establishment recommendations suggesting that we not consume collagen. However, I identified the probable flaw in their conclusions — and used scientific studies to show that humans: cannot convert enough collagen in their own bodies; can consume collagen without any risks; can digest cooked collagen as well as any other animal protein; can assimilate collagen into various tissues; can maintain nitrogen balance; and can provide numerous benefits not only to our connective tissues but also, possibly, to our major systems such as nervous and cardiovascular systems. Thus I reconciled the discrepancies.

But this raises the question: what percentage of our protein should come from completes and collagen? I estimated 70 to 90% completes and 30 to 10% collagens, because we ourselves contain these proteins in these approximate percentages; our ancestors consumed these proteins in these percentages; and scientific studies provide some evidence to support these percentages. For various reasons too complicated to explain here, I surmise we need less collagen when consuming dairy and eggs and more when consuming muscle meats.

So you should consume about 10 to 16% of your calories from protein (80 to 90% of that as completes) and 20 to 30% as collagen. For completes you should consume eggs, dairy, soy, meats, as well as combinations of grains and legumes that are properly processed to improve digestibility and reduce anti-nutrients, toxins and storage proteins. If concerned about generating too much uric acid, leading to inflammation and gout, you should consider consuming more dairy and eggs relative to flesh or even grains and beans.

For collagen you should consume broth, skins, or sausages or hamburgers that are high in collagen; our ancestors additionally consumed highly collagenous tissues like snouts, feet, tails and the like. You should possibly, too, consider consuming gelatin and peptides—even though these are over-processed foods. Additionally, you should carefully process and cook these foods to optimize their digestion. If you want to eat Vegan or Vegetarian entirely, and not consume any collagen, you should include regular amounts of vitamin C in the diet from fresh fruits and vegetables—which helps in the making of collagen inside our bodies.

For our theoretical consumer, at 150 pounds and reasonably fit and active, this equates to:

- 1.0 g equals 68 g of protein (10%)

- 1.2 g equals 81 g of protein (12%)

- 1.3 g equals 88 g of protein (13%)

- 1.4 g equals 95 g of protein (15%)

- 1.5 g equals 102 g of protein (16%) (about the equivalent of 17 eggs per day)

- 1.6 g equals 108 g of protein (175)

At 1.5 g, specifically, this equates to 102 g of protein, anywhere from 80 to 90 g of complete proteins and 20 to 30 g from collagen protein. However, as before, I estimate that when you balance your proteins in this way, you need less protein overall — so these numbers are probably much lower by maybe as much as 30%.

CHAPTER 24

PREMISE 6.1 – OPTIMUM AMOUNT AND RATIOS OF CARBS

As we know, we want to use protein for anabolism to build our bodies while avoiding using it for catabolism. But we use carbs and fats for catabolism (while also using smaller or even trace amounts of them for anabolism) so they are the primary fuels for generating energy or ATP in our bodies. However, they do not necessarily provide energy in the same ways or amounts, so in these sections ahead, I will determine the advantages and disadvantages of each one—which will ultimately determine their ratios in our diet.

I am discussing the macronutrients in the order of their importance in our diet—with protein first, followed by carbs and then fat. Protein is most important because you do not store much of it in your body, so if not receiving enough in your diet every day, you start to break down your muscles. If not receiving enough, you are also prone to overeat carbs and fats to attempt to attain enough protein—which is one of the main reasons moderns become overweight.

Carbs are second most important also because we only store enough of them, as glycogen, to last about one day so, after that, we start glucogenesis and ketogenesis—which we want to avoid. Interestingly, we evolved to store limited amounts of glycogen because it must bind with water, making it heavy. Meanwhile fat is lighter and thus we store enough to last weeks or months. While fat is more essential to our diet over the long-term, it is nonetheless less pressing from day to day.

NUTRITIONAL PHYSIOLOGY

Plants photosynthesize the three sugars, glucose, fructose and galactose, and bond them together to make carbohydrates. After eating them, we digest them into their sugars again and absorb them into our blood. At our live fructose and galactose convert into glucose as soon as possible because these sugars are toxic and our cells cannot metabolize them. Sometimes fructose, in particular, is converts direct into the fatty acid, palmitate. Meanwhile glucose flows through the liver to our cells to make ATP. Our blood only holds about 4 g of sugar at any one time,[236] about the equivalent of one teaspoon of honey even as we might consume 100 g or more in any one meal. So we must absorb sugars gradually while metabolizing or storing them as soon as possible to prevent making our blood toxic.

Glucose is catabolized principally by our brain and red blood cells, as well as some other tissues in our body—but our brain also catabolizes acetic acid and ketones. In the presence of oxygen, one molecule of glucose generates 34 ATP. In the absence of oxygen, glucose ferments to generate small amounts of ATP as well as lactate which is then further catabolized.

On average our brain catabolizes about 130 g of glucose per day,[237] and our red blood cells about 20 g. Some other tissues catabolize minor amounts of glucose, including parts of the kidney and eye, totaling about 30 g per day, so on average humans need at least 180 g of glucose per day—about the equivalent of six apples or slices of bread. However, under certain conditions, other tissues possibly catabolize glucose, including our white blood cells but only about 20 g more. Furthermore, through glucogenesis, our body produces about 20 g of carbs per ay from glycerol, propionate and amino acids—but probably under 20 g per day. So altogether, we likely just need somewhere between 160 to 180 g of carbs per day.

Whether we are napping or writing this book, our brain catabolizes glucose at the same rate;[238] however, dependent upon the task, some parts of our brains are working harder than others;[239] and accordingly some parts of our brains hit exhaustion while other parts are rested and ready—which is why writing and ping pong are such excellent mates. At the same time we cannot increase our mental performance by consuming more sugar than what is needed.[240] But this is not true for our muscles: as they work harder, they catabolize more calories.

Other tissues that prefer fatty acids, such as our muscles and organs, absorb glucose after meals, if only to lower our blood sugar—but then revert back to fatty acids as soon as possible. Our muscles also switch to preferring glucose at higher levels of exertion such as when running because glucose creates ATP faster—but even then, we only need at the most about 60 more g of glucose per hour. When under-consuming carbs, we merely perform glucogenesis and keto-genesis. When over-consuming carbs, especially over-processed ones, we suffer various detriments over the long term, leading to the diseases of civilization.

GLYCEMIC INDEX

While all carbs undergo the same digestion and absorption as mentioned above, some carbs nonetheless digest and absorb into the bloodstream at different rates as compared to others—which in turn determines how much they raise and then lower our blood sugar—which scientists measure through something called the Glycemic Index and the Glycemic Load. When low on the Glycemic Index or Rate, foods digest and absorb more slowly into the blood and in turn only create slight increases and decreases in our blood sugar. When high, foods digest and absorb faster into the blood, causing spikes and troughs. Generally, we want to keep our blood sugar stable at about 5 to 6 mmol per liter of blood both before, during and after meals.[241] However, high glycemic foods possibly fluctuate our blood sugar up or down by 3 mmol, so that our blood sugar is fluctuating between 4.5 to 7.5 mmol, which makes blood sugar both too low and too high — and both of those are problematic. Generally, humans should never increase their blood sugar over 7.1 mmol per liter of blood.[242]

To measure the glycemic index, scientists feed humans 50 g of pure glucose and then measure the response on their blood sugar and then score glucose 100 on their index. Scientists then compare all other foods to this score and then assign that food its own relative score, somewhere under 100. Low glycemic foods are scored between 0 and 55, moderate between 55 and 70, and high when between 70 and 100. However, this form of measurement is not perfect because it does not account for the bulk of any particular food. For example, watermelon scores "high GI" at 76, but to consume 50 g of carbs in watermelon, we must consume about five cups — which is more than what people typically consume. So watermelon is technically high in GI, but in reality, not necessarily when consumed in normal amounts. Meanwhile one donut also scores about 76 and we usually consume two or three of them at once. So in every way, donuts, as compared to watermelon, are both technically and actually high in GI — and given that humans consume so many of them, they are thus problematic as food. Because of this flaw in the methodology, scientists then devised the Glycemic Load — which basically takes the Glycemic Index and then multiplies that by the number of g of sugar in any typical serving of food and then divides by 100 — which then renders more realistic results. I refer to both of these measurements somewhat synonymously from here on out.

Generally, we tend to think that refined carbs, like white flours and sugars, have a high GI, and we tend to think that whole carbs have a low GI. And this is mostly true, but not absolutely true because GI is affected by many other factors so that some refined carbs have lower GI and some whole carbs have higher. For example, both white bread and wheat bread, oddly, score about the same (although this is misleading because we probably would never consume as much whole wheat as white bread because it's bulkier in our gut). Soft drinks do not have an especially high GI because their carb, sucrose, is half fructose, which absorbs into our blood more slowly. Additionally, wheat pastas do not score very high on the index either, because in the process of making pasta, the starch becomes more embedded in the protein—thus slowing down the digestion. At the same time, whole carbs do not necessarily score low in GI—for example, instant oats score high in GI, probably because they are so thoroughly cooked, while rolled oats score much lower. So in other words, there are correlations between refined carbs and high GI; however, the correlation is loose and does not apply in some cases. But keep in mind that GI is merely one way, and not the best way, to measure the overall benefits of carbs.

Overall, carbs' effect on glycemia are determined by the combination of many variables. First, refined carbs do in general score higher—just not always. Second, as you might guess, carbs score higher when more predigested—that is, ground into smaller particles and cooked; however, and oddly, sprouting and fermenting causes grains to score lower in glycemia. Third, carbs tend to score higher when consisting of fewer chains of sugars, such as with sucrose, maltose and lactose, but score lower when consisting of many chains of sugars, as with starch, typically referred to as "complex carbohydrates."[243] Fourth, foods score higher when their sugars are less embedded into other compounds in the foods, including proteins and fats and score higher when less embedded. Fifth, foods score higher when containing lesser amounts of organic acids (the sour and tart compounds found abundantly in fruits, vegetables and fermented foods). Sixth, carbs score higher when consumed with lesser amounts of proteins and fats. Seventh, carbs score higher when containing greater amounts of fructose and galactose, relative to glucose, because these sugars are taken up by the liver for conversion. And lastly, fruits in particular score higher in GI when riper.[244,245,246]

COMPARING CARBS TO FATS

How do carbs compare to fats along several lines of questioning? Which ones provide greater diversity of nutrients? Which ones are easier to digest? Which ones produce more oxidation and glycation or AGES? Which ones generate the most ATP? I will explore these questions — even though the answers are not all that helpful.

First of all, carbs do not necessarily provide more nutrients as compared to fats. While some carbs, like whole fruits, tubers and grains, come attached to many other nutrients, so do nuts and avocados. While we eat many of our carbs with many of their other nutrients removed, we also consume many of our fats that way as well as oils and tallows — and even then refined carbs provide some nutrients and rendered fats provide fat soluble vitamins and phytonutrients. So in the end, this probably does not matter all that much, except to say that too much refined carbs and rendered fats are potentially problematic because both are low on other nutrients.

Second, carbs, or whole carbs, require considerably more digestive metabolism as explained in pior sections as compared to fats. However, fats require many more chemical steps to produce ATP. Third, carbs generate more free radicals, otherwise known as ROS, as compared to fats — but not enough to matter neccesarily. When both glucose and fatty acids pass through aerobic catabolism (the Krebs cycle) to generate ATP, they release various free radicals in the process — which are unstable molecules that can damage other molecules. However, these free radicals also provide benefits by helping to break down various damaged or decaying molecules or cells in our body, while also doing the same to pathogens. Generally, carbs generate the most amount of oxidation, followed by fats and then protein;[247] however the differences are not that significant.

Fourth, carbs do not necessarily contain more antioxidants than fats. Both whole carbs and nuts come attached to antioxidants. However, most refined carbs contain few or no antioxidants — and thus both cause oxidation and provide little protection from it; and the same applies to many refined oils treated with chemicals and the like.[248] But many rendered oils and fats are rich in anti-oxidants, including many plant fats, from olive, palm, sesame and etc, as well as animal fats from butter, egg yolk, and tallow from pastured animals. Meanwhile, some fatty acids, specifically the polyunsaturates, and especially the omega-6, linoleic acid, do not generate more free radicals—but are more likely damaged by them—whereas other saturates and monounsaturates are more stable, as well as glucose. So in the end, carbs do not necessarily contain more anti-oxidants than fats but we probably benefit by eating both carbs and fats that are high in anti-oxidants.

Fifth, carbs generate more AGES (Advanced Glycation End-products) than fats because sugars are the base of these compounds. We typically form AGES when we consume too many carbs and raise our blood sugar, which then causes the sugars, and especially fructose and galactose, to stick to fats and proteins and form damaged molecules that are difficult to eliminate—and thus accumulate in our bodies over our lifetime.[249] However, when consuming carbs in their more natural state as whole fruits, grains and tubers, we are less likely to form AGES or free radicals.

Sixth, carbs likely both generate and sustain more infections in the gut, as well as the body at large, because microorganisms feed way more on sugars than fats.[250]

So overall, I do not think there is much reason to consume carbs over fats, or the reverse, based on these criteria. However, as I will soon explain, we should consume one over the other for other reasons: namely, some produce greater or lesser amounts of ATP at different rates and some tissues prefer one over the other.

ANCESTRAL CONSUMPTION

Even back seven million years ago, our theoretical ancestor already possessed one of the largest brains on earth, which thus catabolized more glucose than other animals. Accordingly, they evolved in jungles rich with fruit that, in turn, provided glucose nearly year-round along with many other nutrients, including potassium, fiber and especially soluble fiber, beta carotene, vitamin C and K, as well as phytonutrients. Also their fruit was probably about 30 to 40% fructose—which delays the flow of glucose into the blood. In eating carbs in this way, our "common ancestor" kept their blood sugar balanced and thus avoided all detriments already mentioned. While evolving further into Australopithecus, then Homo erectus, our ancestors continued to consume fruit while adding more starch as tubers, sedges and grains—and thus consumed less fructose relative to glucose, allowing them to reduce conversions. They also started to predigest their carbs more, allowing us to eat more of them. But they also started to consume more dietary fat relative to carbohydrates, mostly from animals.

As foragers, humans generally consumed considerable amounts of carbs, mostly as starch but also as fruit, berries, and honey. However, they also varied considerably, dependent upon their ecosystem. In tropical environments, they tended to consume more carbs and less starch relative to fruit. In temperate environments, they tended to consume more carbs and more starch relative to fruit. In cold and more barren environments, they tended to consume fewer carbs overall and, in many cases, consumed more protein and fat to compensate for the reduction in carbs.

As agriculturalists, humans consumed more carbs and way more starch relative to fruit, specializing in whole grains, legumes and tubers while fruit was consumed seasonally or, otherwise, fermented so that most of the carbs were removed. Many, too, were not able to access ample animal foods or plant fats and thus consumed enormous amounts of carbs as bread, porridges and various types of beer. As moderns, humans consume even more carbs, as over-processed starches, fruits and sweeteners trend us towards the diseases of civilization.

Overall, our ancestors always consumed carbs, but various amounts of them, depending on their availability in their ecosystems. On the low end they consumed about 50 g per day; but most consumed about 150 to 200 g, and some over that. Starting with Australopithecus through human agriculturalists, our ancestors primarily consumed starches, secondarily fruits and then sweeteners in small amounts—so that they consumed around 60 to 90% glucose and the remainder as fructose or galactose from dairy. Nearly all of their carbs came from whole plants with their fibers intact—even as they rigorously predigested and processed them—and consumed them with protein, fats, organic acids and other compounds that slowed their absorption into our blood. In eating carbs in this way, they experienced all the benefits without any of the detriments.

ESTABLISHMENT RECOMMENDATIONS

For their minimal intake, the nutritional establishment recommends on average about 130 g per day to provide the brain with adequate amounts of glucose—even as we need about 20 more g for our red blood cells. For some reason, they do not set any upper limit at all—even though carbs, even whole carbs, probably become detrimental at higher levels of intake. However, they generally recommend that we consume about 45 to 65% of our calories from carbs—which for our "theoretical consumer" amounts to an enormous amount per day: 270 to 390 g, which equates to between twelve and fifteen pieces of whole wheat bread. They recommend that we consume most of these carbs from whole grains, tubers, and fruits while keeping added sugars, such as sweeteners and juices, at below 25% of total carbs—which tends to keep our intake of fructose lower relative to glucose. However, they do not recommend that we process our carbs, even our whole grains, in any particular way.

SCIENTIFIC STUDIES

Scientists have conducted many studies, mostly indirectly, on determining the proper amount and type of carbs we need in the diet relative to fat. They do this mostly by studying the correlations between over-processed carbs and disease; the relationship between high glycemic carbs and disease; as well as the benefits and detriments of eating diets either high in carbs (and thus low in fat) or low in carbs (and thus high in fat). Oddly enough, but not surprisingly, they do not conduct many studies attempting to find the best ratio of carbs to fats—even as we have at least some studies on this.

OVER-PROCESSED CARBS

We already know from nutritional physiology how over-processed carbs affect us. To reiterate: since these carbs are more compact, we eat too many of them, and then they move through the gut too slowly, trending us towards constipation, but digest too fast, creating spikes in both blood sugar and insulin, so that we become hungry sooner and then overeat again. At the same they potentially raise our blood sugar from the norm, generating more glycation and oxidation that harm our tissues, especially since these carbs are also low in antioxidants. Since we cannot catabolize all the glucose, we then convert it into palmitate—thus making ourselves fatter. Additionally, we lose many nutrients associated with fiber, including the short-chain fatty acids, minerals, B vitamins, phytonutrients and antioxidants, including molecular hydrogen.

Not surprisingly, studies show that diets high in these carbs deteriorate our health over time. On average modern American adults consume about twenty-four teaspoons of sugar per day—or almost

400 calories—or about 15 to 20% of their total calories. In this case, they are 40% more likely to die from cardiovascular disease, compared to those who consume 8% of their calories from sugar.[251] In meta-analysis scientists showed that when consuming higher intakes of added sugars, humans raise triglycerides, total cholesterol, blood pressure and other risks for cardiovascular disease.[252]

When consuming more sucrose and fructose, humans are more likely to have gallstones.[253] When consuming greater amounts of sweet beverages, women are more likely to develop obesity and type 2 diabetes—when consuming over one beverage per day, their risk of type 2 diabetes increased by 80%.[254] When consuming more than two beverages sweetened with sugar per day, women increase their chances of developing coronary heart disease by 35% as compared to those who consume less than one. When reducing the consumption of these beverages, women lowered both their systolic and diastolic blood pressure.[255] However, when consuming whole fruit, humans are at lower risk for type 2 diabetes, cardiovascular disease, cancer, and mortality from any cause.[256] When consuming higher amounts of over-refined carbs, humans increase their risk for mortality and heart attack and strokes.[257]

When consuming the equivalent of 350 g of carbs per day from refined grains, as compared to 50 g, humans increased their risk of early death by 27% and increased the risk of heart disease by 33% and stroke by 47%.[258] When consuming more white rice, humans increase their risk for type 2 diabetes.[259] White rice is associated with an increased risk of type 2 diabetes.[260]

In meta-analysis, scientists showed that, when consuming more whole grains, humans reduce their total cholesterol, as well as bad cholesterol; they also reduce their percentage of body fat and improve their blood sugar after meals and stabilize their utilization of glucose. They also reduce their chances of type 2 diabetes, heart disease, coronary heart disease, ischemic stroke, as well as several cancers and death from all causes.[261] When consuming breakfast made from white bread and sugary cereals, humans are more susceptible to diabetes, heart disease and cancer.[262] When consuming over-processed carbs, humans are more likely to develop fat around the stomach over the course of five years.[263]

When consuming these types of carbs, humans affect their brains, too, specifically their mood and cognition. When consuming more refined carbs, humans are more likely to experience fatigue as well as lower mood and cognition,[264] as well as neurological, cognitive and behavioral impairments, especially related to memory.[265]

When consuming diets higher in sugars (sucrose, fructose), humans may experience brain impairments through oxidative and AGES due to higher blood sugar and insulin resistance. Furthermore, glycation, one form of AGES, facilitates changes in the brain associated with Alzheimer's.[266] After developing type 2 diabetes and cardiovascular disease, which results in part from the consumption of refined carbs, humans are more likely to experience more aeging of the brain, dementia as well as impaired cognition.

And once developing Alzheimer's, humans show the same states in their brain as diabetics—that is, insulin resistance and the slowed catabolism of glucose.[267]

Whole populations are less damaged by the amount of carbohydrate and more damaged by the type of carbohydrate. Over-processed carbs are associated with increased risk of chronic disease; whereas whole carbs are associated with reduced risk—all of which relates to how these carbs affect sugar and insulin in the blood after meals.[268,269]

HIGH GLYCEMIC CARBS

In the studies above, scientists measured the effects of over-processed carbs. In almost all cases these carbs are also higher in glycemia as compared to whole carbs—although the differences, in some cases, are minimal. Based on epidemiological studies, scientists conclude that when consuming high glycemia carbs, humans are inclined towards risk of type 2 diabetes, cardiovascular disease, and, more recently, age-related macular degeneration (AMD) in people without diabetes.[270] When consuming carbs with high GI, humans are more inclined to cardiovascular disease and death.[271] When consuming foods with the highest GI, as compared to foods with the lowest GI, humans are 25% more inclined towards cardiovascular disease, nonfatal MI, stroke, heart failure and death from any cause.[272]

When overweight or obese, humans, when consuming carbs high in GL, were more inclined towards cardiovascular disease and stroke in particular. When consuming carbs high in GI, humans are still inclined towards cardiovascular disease, but less so, but not towards stroke at all.[273] Based on their review of all forms of evidence on GI and GR, one group of experts concluded that, when consuming diets low in GI and GL, humans were less inclined towards diabetes, coronary heart disease, obesity and some cancers, especially in those with diabetes or insulin resistance.[274] When consuming diets high in GL, even in modest amounts, women increase their risk of cardiovascular disease, especially if they are overweight.[275] When consuming low GI foods humans reduced their glycated hemoglobin, fasting glucose, body mass index, total cholesterol, and low-density lipoproteins, but did not change their fasting insulin, insulin resistance, high-density lipoproteins, triglycerides or insulin requirements.[276] Potatoes, which are high glycemic, are associated in all cases with greater weight, hypertension and greater risk of type 2 diabetes.[277]

However, and confusingly, scientists, in other studies, do not show any correlation between high GI and increases in chronic diseases or other pathologies. In one meta-analysis, scientists could not find any correlation between glycemia and disease. However, this study in particular focused on GI which, as noted, is more problematic than GL.[278] In another study, scientists could not find any correlation between GI and GR and mortality from any cause or causes due to cardiovascular disease in men; however, oddly they found the reverse in women.[279] However, these contradictions are probably explained by the design behind the studies and the various factors that influence glycemia. And in any case, the preponderance of the evidence suggests that foods high in glycemia are detrimental; when over-processed they are even more detrimental; when not over-processed they are less detrimental.

WHOLE CARBS, LOW FAT

On diets high in whole carbs, but low in fats, humans typically consume over 250 to 350 g of carbs per day as whole grains, tubers and fruits, usually in conjunction with beans, nuts and other plant foods and smaller amounts of animal foods. In the breakdown of the macronutrients, they consume about 10 to 16% protein, 70 to 80% carbohydrate and 10 to 20% fat; this amount of carbs equates to about twenty slices of thick, whole grain bread. In comparison to the SAD (Standard American Diet), these diets provide great benefits. On these diets humans reverse all the detriments of the refined diets and turn them into benefits. Additionally, they do not overeat proteins. And while low in fat, these diets are not necessarily deficient in fats, or at least not deficient in the essential fatty acids—and otherwise they are rich in minerals, vitamins, phytonutrients and many other nutrients.

However, these diets provide our bodies with more carbs than needed—so we either just convert the excess glucose into fat or force our muscles and organs to catabolize the glucose. Furthermore, these diets are not necessarily appetizing, with too much fiber and secondary compounds while requiring considerable amounts of digestive metabolism.

Some foragers and many agriculturalists consumed close to this sort of diet—even as they always consumed at least 20 to 30% of their calories from fat. These foragers include the Hadza, Pygmies, Natufians, Aborigines, and many jungle cultures who are generally regarded as healthy. Agriculturalists, too, are generally regarded as healthy as well, as compared to most other moderns; they live longer and freer from the diseases of civilization.

The Okinawans inhabit beautiful, subtropical islands south of mainland Japan. They consume about 10% fewer calories than other Japanese, with about 80% of those calories from carbs; about 10% from fat; and 8% from protein. For protein they eat whole fish and pigs as well as broth—and some legumes, and soy most of all. For carbs they consume enormous purple, sweet potatoes, which are low in GI, as well as small amounts of brown rice. For fats they consume small amounts of oils and animal fats. For phytonutrients they consume green and yellow vegetables, ginger, turmeric, bitter melon, and various seaweeds like Kombu which are rich in iodine.

At the time of the studies—generally in the 1950s—the Okinawans were way more active as compared to Americans and lived in more permanent, supportive and interactive cultures while receiving plenty of sunlight. As compared to other Japanese, and other people eating SAD, these people lived longer; as compared to Americans, they were anywhere from around three to eight times less likely to die from coronary artery disease, colon, prostate or breast cancer. They also did not gain weight as they aged. However, women also suffered from delayed menstruation and lactation associated with lower amounts of calories.[280,281,282]

The Tsimane of South America eat similar diets with the same results. They live in the rainforests of Bolivia and both farm plants as well as hunt animals. They consume 72% carbohydrates, 14% protein and 14% fat (38 g per day and 11 g of saturated). They eat all parts of wild game and fish with whole rice, plantain, manioc, corn, nuts and fruits—which they grow themselves. As compared to moderns, they have low blood pressure, blood sugar and cholesterol, including "bad" cholesterol. They are additionally much leaner than Americans and live generally longer as well. Like the Okinawans, they are more active compared to moderns, moving somewhere between four to seven hours per day—and live also in more supportive cultures. As compared to many other cultures, these people have less atherosclerosis than any other people studied in recent times (five times less than Americans); they also have low blood sugar, low blood pressure, low cholesterol, including the "bad" cholesterol, and low weight. However, and paradoxically, they also score high overall in inflammation due to infections, possibly from living in tropical environments. Scientists do not have data on their lifespan or the incidence of other diseases.[283]

Scientists have conducted many other studies on cultures that eat these sorts of diets—and found similar results. The Mediterranean and Blue Zones Diet, which receive lots of attention these days, are somewhat like these High Carb Diets because they include a high diversity of plants. But they are also generally higher in fat, and mostly monounsaturated fat, at 20 to 35% of calories. On these diets, people generally show improvement in their health as compared to most moderns. All of these studies are problematic, however, because they cannot necessarily determine direct causation—that is, these cultures are probably healthier due to their diet but also their lifestyle or, more likely, combinations of all of them.

Many moderns now eat "High Carb Diets" by choice, mostly with the hope of improving upon their health—or more specifically, to reduce the expression of the diseases of civilization. Humans first started adapting these diets because of the "lipid hypothesis" proposed many decades ago—that is, the problematic notion that heart disease is caused by diets high in cholesterol, fat and specifically saturated fat. To avoid the supposed detriments of fats, some thought they should consume enormous amounts of carbs—but in this case, whole carbs. The Pritikin diet encourages the consumption of whole grains, whole fruits, as well as vegetables and small amounts of lean meat, fish and dairy—while eschewing the use of any added oils or fats.[284] These diets are at least 70% carbs, somewhere around 10 to 16% protein, and 10% fat and full of whole plant foods rich in fiber and phytonutrients. The Ornish Diet is similar, encouraging the same foods, while allowing dairy and eggs but not any meat.[285,286]

All these diets also emphasize exercise as a remedy for stress. In any case, they are similar to the Okinawan and Tsimane diets and presumably many others through the ages. Other diets are offshoots of these, including ones that are vegan, vegetarian, or the most recent trend, plant-based—although all these diets, too, may contain greater or lesser percentages of carbs.

Accordingly these diets also show benefits, especially as compared to SAD. To measure the efficacy of these diets, scientists generally use two techniques: they compare them to some version of our modern diet (typically defined as SAD: the standard American Diet) or any random diet that Americans consume; or scientists recruit people with various expressions of the diseases of civilization and then feed them this diet to see if they improve. And with these methodologies, people on these diets tend to lower their weight, blood sugar, blood pressure, insulin, and cholesterol, especially bad cholesterol. Additionally, the Ornish diet shows evidence of improvements in telomerase, telomeres (compounds associated with longevity and reduction in disease), gene expression, prostate cancer, diabetes, blood pressure, depression, etc. So these diets reduce disease and improve health in people who were consuming SAD before or suffering from disease. However, this does not mean that these diets are optimum for humans in general.

LOW CARB DIETS

"Low Carb Diets" are of course low in carbs but high in fats—but vary in their amount of protein from 10 to 30% of calories. When containing higher amounts of protein, in the 30% range, they are typically called "high protein" diets, such as the Atkins, Zone, and even Paleo diets. When lower in protein, and really low in carbs, they are called Keto diets.

When applying the Keto diet to "our theoretical consumer," I derive the following: 2500 calories total; 14% protein; 8 to 20% carbs; and about 70% fat—which equates to about two pieces of fruit per day but 14 tablespoons of fat. However, most people cannot stomach that much fat so they just eat fewer calories overall.[287] Additionally these diets do not specify any type of protein or fat while also encouraging some key nutrients like vitamins, minerals, fiber and phytonutrients.

Since humans only consume about 50 g of carbs per day on this diet, they must generate carbs and ketones to provide the equivalent of that 180 or so g of glucose per day. But on keto diets, people do not consume much protein so they make glucose from fat, specifically from glycerol. And then they make considerably more ketones from fatty acids—which is one of the more complex and prolonged forms of conversion and transport in our bodies. So in short these diets require an enormous amount of conversion. Additionally, these diets are low in fiber and tend to cause more constipation, fewer defecations, as well as lower counts of beneficial bacteria in the colon, which then reduces the production of short-chain fatty acids. However, these detriments are reduced or perhaps nulled if considerable amounts of vegetables, beans, and fruits are consumed.

Some foragers consumed diets similar to this—although in almost all cases, they generally consumed more protein and more carbs. As far as I know, the Inuit Diet is the most ketogenic, but they consumed about 25% of their calories from protein and usually consumed about 50 g of carbs per day. Some scientists have even claimed their diet is not ketogenic. In some cases, herders also came close to consuming this diet but consumed at least 50 g per day from lactose and usually supplemented their diet with small amounts of carbs and also consumed higher amounts of protein. So in short ketogenic diets were not really practiced by our ancestors, unless they could not find enough carbs or protein in their diet, such as during drought.

Although many people claim to receive benefits from these diets, studies suggest more mixed or ambivalent results. Since they are both reducing their consumption and conversion of sugar, people on these diets naturally stabilize their blood sugar, and lower their insulin as well as their insulin resistance; and this in turn helps people lose weight. Additionally, some people propose that the diet helps with type 2 diabetes and other metabolic conditions usually associated with the "diseases of civilization." Paradoxically, the diet also lowers the amount of fat in the blood even while people are consuming so much fat; this happens because in the absence of sugar, our cells uptake and catabolize fats much faster—thus pulling the fat from our blood.

Furthermore, some studies suggest that these diets lower homocysteine—one compound associated with heart disease. Other studies suggest that they lower bad cholesterol while also reducing the size of bad cholesterol and reducing plaques in the arteries; however, this is probably more dependent on the type of fat rather than the amount of it.[288] Some people proclaim the diet provides benefits in overall health and mental and physical performance; however these claims are not substantiated.

In mice these diets produced considerable benefits, including greater longevity, motor function (strength and coordination), memory, and decreased inflammation, decreased tumors and added 13% more to their lifetime—but this was compared to high-carb, not balanced diets.[289] On these diets, about one half of epileptics reduce their seizures by half—and even maintain those benefits once off the diet.[290] When eating these diets to control seizures over long periods of time, children show a risk of slow or stunted growth, bone fractures, and kidney stones, as well as reductions in "insulin-like growth factor 1," which enhances growth.[291]

On these diets people tend to lose weight initially, but probably because they lose their appetite from eating so much fat. But they cannot adhere to the diet, so they gain their weight back. Generally, studies suggest that any diet with the same number of calories performs equally in reducing weight in the short term;[292] but some people adhere to some diets better than others and thus reduce their weight more permanently.[293] However, one study shows that the diet reduced weight over years and also provided other benefits without any side effects.[294]

These diets, while providing some benefits, also generate detriments, including low blood pressure, kidney stones, constipation, nutrient deficiencies and increased risk of heart disease. Additionally, they probably enlarge livers and kidneys—to create extra space for the glucogenesis and ketogenesis—while also increasing the chance of acidosis, slowed growth and increased cholesterol. On these diets, humans reduce their performance during higher levels of exertion because their muscles cannot get enough glucose. They also possibly reduce their mental performance for the same reason—especially if removing carbs from their diet altogether.[295,296,297] On keto diets, mice showed impaired embryonic development, alterations in organ and brain development and structure;[298] however, in general, mice are not as evolved as humans to consume fat.

COMPARISON OF LOW CARB TO HIGH CARB

Fortunately, scientists performed studies comparing different versions of low-carb to high-carb diets. In this study, published in the New England Journal of Medicine, scientists randomly instructed about sixty recruits to consume either high-carb or low-carb diets. On the high-carb diet, recruits were instructed to consume 60% carbs, 25% fat, and 15% protein but not instructed on the type of carbs. On the low-carb diet, recruits were instructed to just limit their intake of carbs to 20 g per day, at least initially, and then eat as much protein and fat as they wished without specifying any type of fat. Both groups met with dieticians about every three months but, otherwise, were not monitored in their intake of food. In the end, the low-carb group lost more weight at three and six months but not at 12 months—which is consistent with other studies mentioned in prior sections of this book, but both claimed to adhere to the diet at about the same rates. The low-carb group generated way more ketones than the high-carb group, as expected, but for some reason, after three months, that was not the case, suggesting the possibility that recruits were not adhering to the low-carb diet. Otherwise, both groups had about the same level of blood pressure, blood sugar, cholesterol, insulin and insulin resistance.[299] Obviously, this study was problematic in too many ways to mention—and in the end, it only appears to suggest that humans on low-carb diets lose more weight as compared to people on high-carb diets—but only in the first three to six months.

In any case, there are many other studies that suggest that low-carb diets are more effective than high-carb diets for losing weight, at least in the short term, while using questionable methods and reaching questionable results.[300] However, another study, which was much better designed, showed that high-carb diets with whole grains, legumes, etc., were more effective than low-carb diets. In this study, all participants consumed the same amount of protein—at 14%—and consumed "minimally refined" foods, meaning whole plants as well as equal amounts of vegetables. On the high-carb diet, people ate 75% carbs and 10% fat. On the high-fat diet, people ate 10% carbs and 75% fat. In the end both parties expressed equal amounts of satisfaction and enjoyment of meals. However, on the high-carb diet with whole carbs, people consumed 600 fewer calories per day by choice than the other group and lost 89 g of weight per day while showing higher levels of glucose and insulin. On the low-carb diet, people lost only 53 g of weight per day.[301] Another study suggests that when people are trying to just maintain their weight, they experience less favorable results on low-carb diets like ther Atkins as compared to whole-carb diets such as the Ornish or South Beach diets.[302]

BALANCED CARBS

As noted before, scientists, unfortunately, do not devote many resources to trying to determine optimum health for humans, preferring instead to focus on disease. However, they have produced at least some studies attempting to show the best balance of carbs and fats for our health. In one study, scientists sent questionnaires to about 130,000 people in eighteen countries about their diet and their health and from this, they discovered that the people who consumed the most carbs, at about 77% of calories, were about 30% more likely to have died as compared to those with the lowest intake of carbs (46% of daily calories). They also discovered that those with the highest intake of fat, at 35% of daily calories, were about 25% less likely to die than those with the lowest intake of fat. Or in other words, higher intake of carbs and lower intake of fat was associated with greater risk of death.[303] In another study scientists found that, amongst 15,000 Americans, those who consumed moderate amounts of carbs, at about 50% of daily intake, died less than those who consumed high or low carbs.[304]

In another study, spanning nearly half a million people from across the globe, especially from third world countries, scientists found that diets high in carbs and low in fats were associated with increased deaths, especially when those diets were heavy in animal proteins and fats.[305] In another study, scientists documented about six thousand deaths mostly from cardiovascular disease, and found that diets higher in carbohydrates were associated with increased mortality and diets higher in fats were associated with less mortality. Furthermore, they could not find any association between any type of fat, saturated, monounsaturated or polyunsaturated, and cardiovascular disease.[306]

RATIO OF SUGAR WITHIN CARBS

As noted before glucose is our principal sugar while galactose are converted in our live to other substances. Additionally, fructose and galactose are more prone to oxidation and glycation; and fructose, in particular, increase the production of uric acid, block the feeling of satiation which causes people to overeat and gain weight. In excess amounts they are associated with various maladies, like fatty liver disease, senescence and other conditions.[307,308,309,310] However, fructose is only problematic in the ways listed above when consumed in high amounts of over-processed carbs, such as fruit juices and sweeteners like table sugar. When consumed in moderate amounts, especially in whole fruit, berries and perhaps even honey, fructose provides enormous advantages, such as helping us uptake and store glucose, balance our blood sugar and insulin and tolerate and metabolize glucose even better.[311,312] At the same time, fruit and berries, in particular, slow the release of all sugars into our blood and provide many additional nutrients not easily found in other foods.[313]

Galactose is also converted into glucose in the liver—even as small amounts are used in the anabolism of our nerve cells. Most foods contain little to no galactose—not enough to warrant any concern, with the exception of dairy, which contains the sugar, lactose— which is half glucose and half galactose. However even if consuming four servings of dairy as milk, yoghurt or kefir, we consume only 16 g of galactose. In high amounts galactose is toxic; however, in these smaller amounts, galactose probably does not provide benefits or detriments.[314]

Our theoretical ancestor consumed principally fruit; however that fruit contained more glucose relative to fructose than our cultivated fruits; so they probably consumed about 60% glucose to 40% fructose. But we then evolved to consume more starch, so that we consumed more like 70 to 90% glucose and the remainder as fructose. During the modern era, however, we started consuming more fructose than even our common ancestor, mostly in the form of sweeteners, like sugar and corn syrup, which are half fructose. The establishment does not make any recommendations regarding the ratio of glucose to fructose or galactose; however, they recommend that humans keep "added sugar" in the form of sweeteners like sugar, honey and juice, to around 30 g, which amounts to about 15 g of fructose per day.

CHAPTER 25

PREMISE 6.2 – FIBER

NUTRITIONAL PHYSIOLOGY

Plants bond sugars together, mostly glucose, to form fiber which serves as the skin and bones of our their bodies, so to speak. While mostly carbs are strings of sugar with one bond between them, fiber forms four bonds, and then stacks these sugars next to each other and then on top of each other. For this reason, fibers are really strong and stable, which makes them either difficult or impossible to digest for most animals.

Scientists categorize two types of fiber: insoluble and soluble. Resistant to water, insoluble fiber binds with our food and slows down the digestion and passage of that food while also binding with cholesterol secreted from our liver into our guts, preventing its uptake into our blood. Once arriving in the colon these fibers soften our stool and then are flushed from our bodies.

Dissolvable by water, soluble fiber also binds with our food, slows down the digestion and absorption of our food, extends satiation while, paradoxically, speeding the passage of our food through our gut. This fiber, too, binds with cholesterol, preventing its uptake into our bodies. Once this fiber arrives in our colon, it is fermented by microorganisms that secrete special enzymes to break down the fiber into sugars and then consume the sugars and release nutrients into our blood, including various vitamins, minerals, phytonutrients, and antioxidants. At the same time, these microorganisms generate wastes that, paradoxically, serve as nutrients for us, including the three short-chain fatty acids, molecular hydrogen, which is the most effective and omniscient antioxidant in the body, and many other nutrients as well.

Our microbiome also generates the three short-chain fatty acids catabolized by many of our cells. Usually, our colons generate acetate the most, followed by butyrate and then propionate—although of course this also depends on our diet. Propionate is transported to our liver and converted into glucose but possibly provides some benefits along the way for anti-inflammation, glucose regulation, weight management and other conditions. Butyrate provides specific fuel for the cells lining our colon while helping those cells connect with each other, which keeps our colon healthy and, in the absence of butyrate, these cells suffer detriments.[315] Additionally, butyrate also helps to prevent cancer and inflammation.[316,317] Acetic acid is catabolized by most cells in our body, perhaps most of all our brains, while also providing other benefits, including greater catabolism of fat, resulting in weight loss, as well as decreased blood sugar, increased insulin sensitivity, and improved cholesterol. Acetic acid also improves the health of our brains, resulting in better concentration, performance and mood while reducing depression and anxiety. Additionally, acetate as well as butyrate are also converted into other fatty acids.[318]

These microbes in our colon, known as the human microbiota, contain ten times more cells than the rest of our body—even while most of those cells are prokaryotes and are thus minuscule compared to our own cells. They weigh about two pounds altogether, about the same as one pineapple, and contain 100 times more genes than our own. Altogether, our colon contains about five hundred different species of these organisms with nearly all of them as bacteria, but some as archaea and prokaryotes like yeast and protozoa as well as the occasional parasites like worms.[319] All these organisms, in turn, are afflicted with viruses. While nature collectively contains about seventy classes of bacteria, our colon contains only about nine of them—and only one of thirteen archaea. The most abundant phyla in our colon, the bacteroidetes, produce acetic and propionic acid, while firmicutes, the next most abundant, produce butyric acid. Outside of that we range enormously with different microorganisms, including proteobacteria, actinobacteria, fusobacteria, cyanobacteria, and verrucomicrobia.[320,321] However, all these organisms also cross feed as well—that is, while digesting one form of fiber, one bacteria generates compounds that other bacteria eat.[322] Bifidobacteria and lactobacilli break down the fiber, fructan, and generate both acetate and lactate; and then other bacteria, including Eubabacterium and others, consume those short-chain fatty acids and generate butyrate.[323]

Generally, several factors influence the amount and diversity of our microbiome. First, when consuming more soluble fiber, we increase the size of our microflora because they have more food to eat; this also lowers the PH of the colon which helps them proliferate even more. Second, when increasing calories in our diet, we tend to reduce the diversity of our microbiome, although for reasons that are not understood.

Third, when eating a greater diversity of whole plants, we increase the diversity of our microflora and the nutrients they create. Plants produce more than one type of fiber and different plants generate different combinations of fiber. Furthermore, our various microorganisms only feed on specific types of fiber—and then, in turn, generate specific types of nutrients such as butyric acid. So when eating greater diversities of plants, we consume greater diversities of fiber—which then feeds greater diversities of microorganisms, which then generate greater diversities of nutrients for ourselves—which in turn provides greater benefits for our health. However, too much fiber might prove detrimental because it may slow digestion, make our meals unpalatable, and overfill our stomachs while preventing us from eating enough of the macronutrients.

Fourth, animal foods, relative to plant foods, affect the microbiome: animal foods promote the growth of some organisms, plant foods the growth of others.[324] When we properly digest our food, only about 10% or less of our protein arrives in our colon; however, some of that protein is beneficial, providing nitrogen, amino acids and other nutrients for our microflora. And in the process of eating those compounds, our microflora in turn releases short-chain fatty acids, just as they do from fiber, but other compounds as well, including branched-chain fatty acids, phenols, amines, sulfides, gases, ammonia and TMAO—some of which are beneficial, and others detrimental, with TMAO linked specifically to heart disease.[325]

Fifth, fatty acids in our diet also influence our microbiome. Polyunsaturates and monounsaturates increase certain kinds of bacteria, including firmicutes, as well as ones that generate lactic acid, and saturates increase others.[326] Sixth, the size of the fiber also influences our microbiome; some species feed on larger pieces, others smaller, so for that reason, how you process and predigest your plants ahead of time, and then chew them, in turn affects your microbiome.[327] Interestingly, our microflora also generates lots of lactate (lactic acid) and alcohol during fermentation. But lactate competes with the three short-chain fatty acids for uptake and for that reason, most lactate is not absorbed into our blood but eaten by various other microorganisms in our colon.[328] Alcohol, too, is generally not absorbed as well but consumed preferentially by various microorganisms.

ANCESTRAL CONSUMPTION AND ESTABLISHMENT RECOMMENDATIONS

While our theoretical ancestor consumed enormous amounts and diversities of fiber, all hominoids, from Australopithecus to foraging humans, consumed less because they consumed fewer plants relative to animals—even though they continued to consume enormous amounts of fiber compared to modern humans. Some carnivorous foragers consumed little to no fiber.

Agriculturalists also consumed enormous amounts of fiber, but with less diversity because they consumed fewer plants, mostly just one or two grains and legumes, and thus reduced the diversity of fiber and the nutrients that come from fiber. Agriculturalists probably consume about equal parts soluble to insoluble fiber—but not in all cases for this reason. Most grains contain higher amounts of insoluble fiber, including wheat, rye, and quinoa, while only oats and barley contain higher amounts of insoluble fiber but most legumes are higher in insoluble fiber. Herders, of course, consumed little to no fiber—but without any obvious detriment. Moderns sharply reduced the amount and diversity of fiber in our diet, trending us towards the diseases of civilization. But overall, across time, most humans consumed somewhere between 15 to 100 g of fiber per day at around equal parts soluble to insoluble fiber due to the diversity of their plants, including fruits, tubers, seeds, leaves and others. The nutritional establishment recommends around 25 to 30 g of fiber per day, with only one fourth of that, 6 to 8 g, coming from soluble fiber—although I do not know the reason for that. They do not set upper or lower limits or even encourage diversity in fiber.

SCIENTIFIC STUDIES

Many studies show the benefits of fiber to our health. Fiber accelerates intestinal transit while increasing satiety and helping with weight management.[329] Several meta-analyses show that when consuming higher intakes of fiber, humans of all ages reduce their risk for developing coronary artery disease, stroke, hypertension, diabetes, obesity, specific cancers, infections, respiratory infections and specific diseases of the gut, including gastroesophageal reflux disease, duodenal ulcer, diverticulitis, constipation, and hemorrhoids. Additionally, they lower inflammation, blood pressure and cholesterol levels and live longer.[330,331,332,333,334]

When obese humans supplement with fiber, they lose weight.[335,336] When diabetics consume diets moderate in carbohydrates but high in fiber, they lower their blood sugar, triglycerides and cholesterol and receive other benefits.[337] Fiber also improves the absorption of vitamins and minerals into the diet, including vitamin K and various forms of vitamin B.[338]

Soluble fiber, in particular, stabilizes blood sugar and insulin sensitivity while providing multiple benefits for the colon: catabolic fuel, inflammation reduction, and better immunity. Additionally, this fiber increases the pH of the colon, which protects from polyps, and increases the absorption of minerals while also increasing the functioning of the immune system, including T helper cells, antibodies, leukocytes, cytokines, and lymph. It also suppresses cholesterol synthesis by the liver and reduces the bad cholesterol and triglycerides responsible for atherosclerosis.[339]

BRAIN

Interestingly, our brains are connected to our colons through something called the GBA, the gut brain axis—which is the connection between the central and enteric nervous systems. These nervous systems communicate with each other through neural, endocrine, immune and humoral links.[340] More specifically, our microbiome releases chemicals, including cytokines, neurotransmitters, neuropeptides, chemokines, amino acids and short-chain fatty acids that then travel to the brain via the blood, nerves and other pathways, and impact the functioning of the brain, specifically the hippocampus, prefrontal cortex and amygdala.[341,342,343] In other words, our microbiome, and its various byproducts, can influence the way we think and feel.

However, to date, most of the research on this subject has been done on mice. When their colons are predominant in Bacteroides fragilis, mice tend towards autism; when that bacteria is reduced, they tend to normalize.[344] When born sterile, without any microbiome, mice show greater anxiety and depression.[345] In another experiment scientists switched the microbiome of timid and bold mice—and found that their behaviors switched as well.[346] In another study scientists took the microbiome from humans with irritable bowel syndrome and then planted that into mice, which caused them to exhibit signs of autism, although this did not happen when the same was done to healthy humans. [347]

In humans certain metabolites from our microbiome promote the production of serotonin in the lining of our gut[348] and other metabolites promote the production of GABA, the neurotransmitter that promotes calmness in our brain.[349] Furthermore, when humans consume less fiber, they are more likely to experience mental disorders, including anxiety, autism, depression and schizophrenia, as well as autism and Parkinson's,[350,351,352] but when given some strains of probiotics for one or two months, some humans experience improvements in various mental health conditions.[353] Additionally irritable bowel syndrome is influenced by our microbiome.

SUMMARY AND RECOMMENDATIONS

Humans obviously can consume different ratios of carbs to fats and survive, and even prosper and reproduce—as witnessed by various cultures and studies across the globe. At the extremes humans range somewhere around 90% carbs to 10% fats of overall calories—and then to 10% carbs to 90% fats. However, most humans have consumed about 30 to 60% carbs of their overall calories, and thus about 50 to 20% fat. But this still asks the question: Is there some optimum ratio between the two? To answer this question, let's summarize our four methodologies.

At the lower end, scientists estimate humans need about 50 g of carbs per day to survive long term, even while performing considerable amounts of gluconeogenesis and ketogenesis. To eliminate those conversions altogether, we need around 180 g for our brain, red blood cells, and parts of our eyes and kidneys. But we also possibly need more or less for various tissues. But we possibly need less dietary glucose because we perform glucogenesis from other molecules like propionate and glycerol even under normal conditions. We also possibly need less because some of these tissues that prefer glucose, such as the brain, also catabolize other nutrients, especially acetic acid from our colon and alcohol (which is converted into acetic acid). So the more fiber and alcohol we consume, the less dietary glucose we need.

But in any case we should consume at least 180 g of glucose per day on average, and possibly some more—but we cannot increase our mental performance by consuming more than that because our brain metabolism is fixed. Additionally we need more glucose for more intense forms of exercise—but only about 50 to 100 per hour of exercise. Our human ancestors followed these same patterns, consuming at least 50 g per day, but normally around 100 to 200 g and even more than that, especially during agriculture. At the lower limit the establishment recommends at least 150 g per day for our brain, but for normal limits, they recommend between 255 to 360 g of carbs per day, which is an enormous amount relative to the other methodologies.

Scientific studies do not show any lasting advantages to consuming diets low in carbs—that is, at somewhere around 20 to 100 g per day. And studies also only show detriments to eating diets high in refined carbs like white flours and sugary drinks.

However, studies show benefits to consuming diets high in whole and diverse carbs as compared to the SAD diet—but that does not mean these diets are optimum for several reasons. First, they contain more carbs than necessary. Secondly, people tend not to like these diets. And thirdly, the benefits of this diet probably come from the addition of other nutrients, including more polyunsaturates, minerals, phytonutrients, anti-oxidants and fiber—and the reduction of meat. Meanwhile several scientific studies show that diets that balance carbs and fats are the most beneficial with carbs somewhere around 30 to 50% of calories.

Nutritional physiology of carbs:

- minimum: 50 g

- optimum: around 180 to 255 g

- add more for exercise over 50% of max

Ancestral Consumption:

- minimum: 50 g

- optimum: around 100 to 255 g (except for some agriculturalists)

- add more for exercise over 50% of max (probably only occasionally reached these levels)

Establishment recommendations:

- minimum: 150 g

- optimum: around 255 to 360 g

- add more for exercise over 50% of max

Scientific studies:

- minimum: 50 g

- optimum: around 180 to 280 g

- add more for exercise over 50% of max

For this diet, I recommend that you consume between 30 to 50% of your overall calories from carbs with 40% around optimum for most people who at least do some exercise.

But how should we consume our sugars within these carbs? We should consume more glucose relative to fructose and galactose. As compared to glucose, fructose and galactose require conversions and are more unstable and potentially damaging in higher amounts. Our ancestors always consumed at least 60% of their sugars as glucose—and in many cases, consumed more like 80 to 90% of their calories from glucose. The establishment obliquely recommends that we limit our intake of fructose by keeping "added sugars" like table sugar and fruit juices to under 10% of calories—which amounts to about 23 g of fructose per day. Scientific studies clearly show the detriments of excess fructose, especially when consumed in table sugar, corn sweeteners and possibly even fruit juices. So this diet recommends somewhere between 70 to 90% of carbs from glucose and the remainder from fructose or galactose—which is easily accomplished.

You should consume your carbs as whole plants with their fibers intact. This accords, without exception, to all four methodologies, especially because refined carbs are the leading cause of the diseases of civilization. And obviously, too, fiber helps regulate the absorption of nutrients while providing additional and beneficial nutrients. For even better results, you should consume mixtures of whole plants—which then provides greater diversities of fiber and, in turn, provides greater diversities of nutrients, including phytonutrients.

When eating carbs in this amount, you provide all the glucose you need for your brain, red blood cells and other tissues in the proper amounts while preventing glucogenesis and ketogenesis. You also avoid all the many detriments associated with overeating carbs, and especially refined carbs. When eating your sugars in these ratios, you provide your body with its preferred sugar, glucose, while providing enough fructose to aid in the metabolism of glucose—but not enough fructose to cause detriments while also receiving the benefits of fruits. When properly processing these carbs, you avoid the detriments of secondary compounds.

In the end, this diet recomends for the carbs that we all recognize as both ancient and satiating—that is, whole grains, whole potatoes, whole legumes, whole fruits and vegetables. For our theoretical consumer, for his 2300-calorie diet, this equates to:

- 30% of calories from carbs: 172 g.

- 40% of calories: 230 g.

- 50% of calories: 287 g – too much.

- 60% of calories: 345 g – too much.

- 70 to 90% glucose.

- 10 to 30% fructose/galactose.

This results in about 20 to 40 g per day, half soluble, half insoluble. If you want to lose weight, you should keep carbs at the lower to middle end of the range, under about 150 to 200 g, to prevent lipogenesis. If you want to delay ageing, you should probably still consume about this same amount of carbs, perhaps on the lower end, because I doubt you want to starve your brain to live longer.

CHAPTER 26

PREMISE 6.3 – OPTIMUM AMOUNT AND RATIOS OF FATS

In the sections above I defined total calories for the optimum diet and then the percentage of those calories from protein and carbs; by default, the remaining calories come from fat (although we can also receive smaller amounts of calories from other nutrients like alcohol and lactate). So even though fats in this case are predetermined, I will nonetheless use this section to explain why we need this amount of fat—and why we need our fatty acids in approximate ratios to each other.

NUTRITIONAL PHYSIOLOGY

Fats consist of smaller nutrients called fatty acids that are shaped like snakes that all have the same heads but different tails that vary based on their degree of saturation, length and curvature. Saturated fats are the most resistant to oxidation and damage, followed by monounsaturates, and then polyunsaturates; the omega-6, linoleic acid, regularly oxidizes into compounds called OXLAMS, some of which are implicated in disease.[354]

We humans regularly consume only about 10 to 15 types of long-chain fatty acids from all classes. However, we receive short-chain fatty acids from our colon and fermented foods, and medium-chain fatty acids and from coconuts and dairy. We can convert saturates and monounsaturates inside our bodies from glucose and glycerol. We cannot make the polyunsaturates but we can convert the shorter into the longer chains — and the reverse.

We receive nearly all fatty acids from our diet from the phospholipids and triglycerides from both plants and animals. We then digest them into individual fatty acids, reconstruct them back into triglycerides, pack them into lipoproteins along with proteins, cholesterol and the fat-soluble vitamins A, D, E and K — and transport them through the blood to various cells. At the cell the lipoproteins are deconstructed into fatty acids, etc, and then absorbed into the cell through several mechanisms.

It appears that most saturated and monounsaturated fats digest and absorb into our blood at about the same rate, at well over 90% — while polyunsaturates possibly somewhat lower than this.[355] Meanwhile, short- and medium-chain fatty acids require little digestion and then travel through the blood without the use of lipoproteins, making them one of the most efficient forms of nutrients.[356] Our blood can only hold about 6 g of fat at any one time, about the equivalent of one-half teaspoon of butter[357] — which is about as minuscule as the amount of glucose. Evidently, we must thus absorb fats slowly and then uptake and catabolize them quickly or otherwise send them back to the liver or store them.

Anabolism and Catabolism

Once fats are absorbed into the cell, only small amounts of them are anabolized to make phospholipids for cell membranes. These fatty acids are all long-chains but from all levels of saturation and shape and our membranes are approximately equal parts saturates, monounsaturates, and polyunsaturates — but this also varies with the type of tissue and the diet. When consuming too much omega-6 and not enough omega-3, we can collect too much omega-6 in our brain, to our detriment.[358]

So nearly all fatty acids are catabolized in the mitochondria to make ATP. Our muscles and organs prefer to catabolize fatty acids under usual conditions whereas our nervous tissues and red blood cells cannot catabolize them at all. Longer-chain fatty acids generate around 100 ATP but more slowly, whereas shorter chains generate around 10 ATP but faster, whereas glucose produces about 30 ATP. However, fatty acids generate about the same amount of ATP by weight—so, as far as scientists know, one form of fatty acids does not generate ATP better than any other.

When the cell is sated, any remaining lipoproteins are transported back to the liver and recycled into new lipoproteins and sent back into the blood. During this stage, some believe that we initiate the process of heart disease. If our cells remain sated, the lipoproteins are then absorbed into our adipose, deconstructed back into fatty acids, then converted into other fatty acids and stored as body fat.

All humans contain about the same ratio of fatty acids in their body fat, regardless of how they consume them in their diet. In other words, if you consume nothing but dairy fat or olive oil, you will nonetheless store your fatty acids in the same ratios: around 30 to 40% saturates, 40 to 50% monounsaturates, and 10 to 20% polyunsaturates, mostly as omega-6 and trace amounts of omega-3, even as the ratios of the omegas does vary more with the diet. This is not too different from how we store fatty acids in our phospholipids.

Scientists do not know why we store fats in these ratios—but perhaps for cushioning, insulation and stability to the body. Or maybe we better catabolize them in these ratios.

When under-consuming fatty acids, we pull them from our body fat without needing too many conversions, so this process is efficient and even optimum for our metabolism. However, we start to slow down our metabolism for self-preservation once reaching about 8% body fat for men, and 15% for women—which is way below the body fat of nearly all modern humans. When under-consuming fats, we are prone to overeating carbs and our muscles and organs uptake more glucose even while preferring fats.

When over-consuming fats without increasing exercise, we possibly trend towards obesity as well as other expressions of the diseases of civilization—but not necessarily. When over-consuming carbs, we trend towards eating more of them as explained before. However, when over-consuming fats, we trends towards satiation; at the same time, we become more efficient at catabolizing them while raising our metabolism. In any case, over-eating fats is not nearly as common or detrimental as over-eating carbs. When over-consuming fats in the wrong ratios, however, we potentially generate more detriments, including heart, liver, digestive and inflammation issues. We also possibly increase the number of conversions of one fatty acid into another in our liver and adipose.

ANCESTRAL CONSUMPTION

From the common ancestor to Homo sapiens, our ancestors evolved towards several trends in the consumption of fats. First, they trended towards consuming more fats relative to carbs. While our common ancestor consumed about 20% fat in their diet, including colonic fats, human foragers and agriculturalists consumed somewhere between 30 and 60% of their calories from fats. Second, they trended towards consuming more animal fats compared to plant fats—so they consumed more saturates and monounsaturates and fewer polyunsaturates. While the common ancestor consumed around 30 to 50% polyunsaturates, human foragers and agriculturalists consumed around 10 to 30% polyunsaturates.

Third, they trended towards more dietary fats relative to colonic fats as they reduced the amount of fiber in their diet. The common ancestor received somewhere around 200 to 300 calories from colonic fats while humans received somewhere between 0 to 200 calories; and this difference is way more extreme if adjusting for size.

Fourth, they trended towards more omega-6 relative to omega-3, from eating more seeds richer in omega-6. While our early ancestors consumed about two parts omega-6 to one part omega-3, some humans consumed ten to one part or even higher. As noted before, it seems self-evident that dietary fat, mostly from animal fat, provided the extra calories needed to sustain our evolution towards bigger bodies and brains.

As humans our ancestors, across cultures, consumed many different amounts of fats and ratios of fatty acids, dependent upon their availability in our ecosystems. When humans inhabited colder environments, they consumed fewer carbohydrates, generated less colonic fats and consumed more dietary fats, mostly from animal fat, that are about equal parts sats and monos. While these fats are low in polys, they compensated by eating brains and eyes that are high in these fats. They sometimes specifically selected for smaller animals, like whole birds, rodents, fish and insects, possibly because they are higher in polyunsaturates. Cro- Magnon Man, in Ice Age Europe, consumed enormous amounts of reindeer fat; the Plains Indians, buffalo and bear fat; the Inuit, seal, whale, and reindeer fat. When inhabiting warmer environments, humans consumed both plant and animal fats—thus eating more monounsaturates comparted to saturates and higher amounts of polyunsaturates with more omega six. These people include the Hadza, Aboriginals, and Natufians.

Agriculturalists continued to consume animal fats as one of the staples in their diets—unless they were unprivileged—from sheep and goats at first, then cows and pigs, and even later eggs and dairy. But they intermixed these animal fats with oily seeds high in the shorter-chain polyunsaturates, including sesame, chia, flax, poppy and nuts as well. They probably did not start to consume plant oils in large amounts until about 5,000 years ago. Herders consumed enormous amounts of both ruminant and dairy fat, without any plant fats—and thus consumed enormous amounts of saturates and lower amounts of monounsaturates and polyunsaturates.

Generally, humans could adapt to these differences in the

amounts and ratios of fatty acids through several mechanisms: First, they could eat more carbs relative to fat—and thus eat more or less fat overall. Second, they could catabolize any fatty acid equally as well as any other, allowing them to consume wide ranges of ratios of fatty acids. They possibly adapted through converting one fatty acid into another, whenever necessary. Or they adapted through epigenetics; or natural selection, as some cultures are possibly more adapted to more fat or different ratios of fatty acids than others.

While humans consumed diverse types, amounts and ratios of fatty acids, they stayed within certain ranges in most, or almost all, cases. Overall, we generally consumed somewhere between 30 to 60% of our calories from fats, with 30 to 50% of that from saturates, 30 to 60% from monounsaturates, and 10 to 30% from polyunsaturates, with somewhere between one to ten parts omega-6 to one part omega-3.

However, our ancestors trended towards the middle of these ranges whenever possible—that is, when those fatty acids were available in their ecosystem. The general patterns look something like this: humans generally strive to consume animal fats from large animals that are high in saturates but then they combine that with other animal fats, mostly from smaller animals, or other plant fats—that are higher in monounsaturates and polyunsaturates.[359] This, in turn, trends them towards the approximate ratios of fatty acids which, interestingly, resemble the same ratios we use to store fat in our bodies in our phospholipids and body fat.

As moderns we radically changed our consumption of fats. First, we consume larger amounts of fats. Second, we refine an enormous amount of seeds into plant oils. And all of this in turn changed our ratios of fatty acids, so that we now consume in some cases less saturated fat, more monounsaturated fat and way more polyunsaturated fat much higher in omega-6 as compared to omega-3. And we combined all of this with greater amounts of carbs—which are then converted into fat.

ESTABLISHMENT RECOMMENDATIONS

Generally the nutritional establishment recommends that we consume 20 to 35% of our calories from fat—and that only 20% of that fat comes from saturated fats, about 60% from monounsaturates (or 53 g per day or 17 g per meal) and that the remainder, about 14 to 30%, come from polyunsaturates—or about 20 g per day or 6 g per meal (at 25%). The establishment does not recommend ratios of omega-6 to -3 but, nonetheless, recommends that we consume about 1 to 2 g of omega-3 per day. On average this results in ratios that are about ten parts omega-6 to one part omega-3; but at the extremes twenty parts to one part.

When applying establishment recommendations to the diet of our theoretical consumer who needs 2400 calories per day, the results are:

TOTAL FAT:
- 25 to 35% of calories from fat.
- 30% equals 720 calories from fat.
- Or 80 g of fat per day.

TOTAL SAT:
- 20% of total fat calories.
- 144 calories per day.
- 16 g per day (about six tablespoons of fat).

TOTAL MONO
- 60% of total fat calories.
- 430 calories per day.
- 48 g per day.

TOTAL POLY

- 20% of total fat calories.

- 240 calories.

- 26 g per day.

Somewhere around ten to twenty parts omega-6 to one part omega-3.

LIPID HYPOTHESIS

Generally, the nutritional establishment recommends we consume only small amounts or percentages of animal or saturated fats—based on the "lipid hypothesis" that associates these fats with heart disease. However, the hypothesis, even though containing some truth, is troubled for many reasons and even remains mostly unproven over the past 100 years, even after scientists tried different variations on the hypothesis. In simpler terms, the lipid hypothesis developed something like this: Back one hundred and fifty years ago, scientists observed damage to human arteries in the form of plaque, which they noticed consisted mostly of oxidized cholesterol, calcium, and other debris from immune cells. After some further deliberations, they thus concluded that cholesterol found in animal fats causes heart disease—but this was never proven and then later disproven.[360]

Furthermore, they concluded that because fats increase cholesterol in our blood when we consume them, then fats in general cause heart disease—but this was never proven either. Additionally, they concluded that only some kinds of fats, most particularly saturated fats, found most abundantly in animal fats and palm oil, cause heart disease—because "saturated fats" cause our own bodies to produce more "bad cholesterol" which then causes heart disease. Meanwhile, they concluded that liquid oils, high in monos and polys, cause our own bodies to produce more "good cholesterol" that prevents heart disease—which in fact removes bad cholesterol from our arteries. The establishment still makes this claim even while the evidence is somewhat mixed.[361,362,363,364] Additionally, they now appear to think that only certain types of bad cholesterol result in heart disease.[365] Scientists are now uncovering past conflicts of interest and bad science related to all of this.[366]

Because of their recommendation that we avoid fats in general and animal fats even more, the nutritional establishment then, some decades ago, recommended that we instead consume more carbohydrates—which in turn spawned the whole "low fat" movement in the food industry, with products that featured little fat but enormous amounts of refined carbohydrates. In turn, this made Americans fatter and trended them more towards the diseases of civilizations, including, paradoxically, heart disease. In the past decade, the establishment recanted on that and stopped their campaign against fats.[367] Now they recommend that, as seen from the presentation above, we eat more fats but mostly as plant oils that usually contain trace amounts of saturates and then high amounts of monounsaturates and polyunsaturates with omegas in varying ratios to each other. In some cases, these oils contain way more omega-6 than omega-3 and only some are balanced.[368]

Nutritional physiology: cholesterol

Similar to phospholipids, cholesterol is used to make our cell membranes while providing different attributes from fatty acids. Cholesterol helps pack membranes more tightly while providing greater stability as well as, paradoxically, greater flexibility, allowing our cells, and especially our muscle cells, to expand and contract and otherwise change form without rupturing. Unlike fatty acids, which change with the temperature, cholesterol remains mostly stable, staying more rigid in heat and more flexible in the cold, helping the membrane maintain homeostasis. Cholesterol also helps our nerves and neurons develop myelin sheaths, which insulate them, and helps electrical impulses travel down them faster. Finally, cholesterol is also the precursor for various hormones, including cortisol, aldosterone, progesterone, estrogen and testosterone, as well as the precursor for vitamin D. Interestingly, plants do not make cholesterol, which in turn makes their cell membranes too delicate so they evolved to form cell walls on the outside of their membranes that are made from sugars.

Humans consume about 20% of their cholesterol in their diet from various animal foods like brains, eyes and eggs, which are rich in cholesterol, and then dietary fats like butter and tallow, which contain smaller amounts. Humans make the remaining 80% in their bodies, mostly in their liver and intestines, from acetyl-CoA. Our brain also makes its own cholesterol which does not pass the blood-brain barrier. When humans consume more dietary cholesterol, we make less of it, so that in the end dietary cholesterol has little, if any, effect upon blood cholesterol.[369]

Once absorbed from our guts into our blood, cholesterol binds with fatty acids and fat-soluble vitamins into lipoproteins for transport to our various cells—which are then absorbed by the cell. But if sated the cell rejects the cholesterol which then causes the lipoproteins to circulate back to our liver, where they are processed into new lipoproteins.

Some of these lipoproteins contain lower amounts of protein and higher amounts of fats, including VLDL (very low density in protein), IDL (intermediate density in protein), and LDL (low density in protein). However one in particular, HDL (high density lipoprotein), contains greater amounts of protein and lower amounts of fats. Furthermore, these lipoproteins contain different types of proteins called apolipoproteins or apoproteins, which in turn latch onto receptors in the cells in different ways, for different purposes, to deliver the lipids into the cell or even close to the cell for uptake.[370]

Once released from the liver, these lipoproteins then travel around the blood to the cells again. LDL carries the most cholesterol to the cells and some cells take the fatty acids from the LDL but leave the cholesterol, making these lipoproteins even thicker in cholesterol.[371] At that point the LDL may circulate back around into the liver for reprocessing and recycling. However, in some cases, the LDL is targeted, oxidized and then consumed by immune cells, called monocytes, which form foam cells that trap themselves amongst damaged endothelial cells inside our arteries, along with other forms of debris. Later they are coated with calcium, forming plaques and atherosclerosis which then causes high blood pressure and heart disease.[372] At some later point, these plaques tend to rupture, leading to heart attacks and strokes. For this reason, LDL is defined as the "bad cholesterol," even though the terminology is misleading because LDL is not cholesterol; it just carries cholesterol.

However, for all of this to initiate, our arteries must first suffer some form of damage, not related to cholesterol or any other fatty substance; instead, our arteries are first damaged in some way by viruses, bacteria, high blood pressure, sugar, and many other factors. This, in turn, triggers inflammation, which in turn causes these monocytes to accumulate at the site, penetrate the linings of the arteries, and then become trapped as macrophages.[373]

Meanwhile, the lipoprotein, HDL, which is lowest in cholesterol, takes up excess cholesterol from our cells and blood. Even better, HDL uptakes cholesterol from foam cells, reversing the damage. They then carry the excess cholesterol back to the liver in the process called "reverse cholesterol transport," where the cholesterol is further oxidized, delivered to the gall bladder, turned into bile which is then secreted into the gut and used in the digestion of fats.[374] At that point, about half of the cholesterol in this bile, about one gram, is then reabsorbed back into the blood which accounts for most of the 20% of the cholesterol that comes from our diet. The other half is then emptied into the colon and fermented into coprostanol, which is then excreted in the feces.[375] If we consume more fiber and other plant substances, we dump more of that cholesterol into our colon — but again this does not affect our blood cholesterol in any way. So for these reasons, HDL is called the "good cholesterol."[376]

So the lipid hypothesis in its current form claims that animal and saturated fats raise bad cholesterol and this, in turn, leads to the formation of plaques and heart disease. Meanwhile monounsaturates and polyunsaturates lower bad cholesterol and raise good cholesterol, preventing the formation of plaques and even possibly reversing some of their development. So we should avoid saturated fats as found in animal fats and instead consume monounsaturated and polyunsaturated fats as found in plant oils.

Problems with the lipid hypothesis

The lipid hypothesis seems troubled for numerous reasons. First of all, cholesterol does not cause the initial damage to the cells that results in atherosclerosis; other factors cause that damage, some known, some unknown. Second, once damaged, these cells cannot heal themselves, but this also is not caused by cholesterol. Tangentially, the hypothesis does not make sense considering the nutritional physiology of cholesterol — that the molecule is not only inherently undamaging, but entirely beneficial, especially for the functioning of our brain. Furthermore, cholesterol is entirely regulated by several mechanisms in our body.

Even if we skip the above evidence and assume that saturated fats cause heart disease by raising bad cholesterol, the lipid hypothesis still does not make much sense. First of all many other factors influence cholesterol, and the ratio of good to bad cholesterol, other than fat or saturated fat. Our ratios of cholesterol are mostly set by genetics, at about 50%, so that we can only control these ratios to some limited extent through diet and lifestyle.[377] Second, we generate more cholesterol overall when merely eating more. Third, refined carbs appear to generate more bad relative to good cholesterol than saturated fat — and much more conclusively, they also raise the total amount of lipoproteins in the blood (because glucose blocks the absorption of fat by cells) while also increasing insulin resistance. Fourth, exercise lowers bad cholesterol. Fifth, statins lower bad cholesterol while not really raising good, as well as the overall amounts of lipoproteins in the blood—which probably disrupts our cholesterol balance, leading to their numerous side effects. So given all of this, both the amount and ratios of cholesterol in our blood are determined mostly by our genetics, and then many factors related to diet and lifestyle, so that, in the end, the amount of saturated fat in your diet affects your cholesterol presumably less than around 30%.

But even if you believe, against the evidence, that saturated fat raises bad cholesterol enough to matter, we encounter even more problems. Animal fats in reality do not raise bad relative to good cholesterol; they just raise total cholesterol and thus raise bad and good cholesterol about equally.[378] This is because these fats, while called saturated, contain in reality about equal amounts of monounsaturates and some contain more, including chicken, duck, rodent and even to some lesser extent pork; and marine fats are equaly parts of all three classes of saturates. So, these animal fats probably lower bad and raise good cholesterol.

And even more weirdly, only three saturates when isolated generate more bad relative to good cholesterol: palmitate, which is abundant in our diet, and laurate and myristic, which are rare in our diet unless we consume lots of coconut oil. Meanwhile the second most abundant saturate in our diet, stearate, does not increase or decrease bad cholesterol, perhaps because 20% of it is usually converted into oleic acid inside our cells. Or in other words, animal fats do not raise bad relative to good cholesterol, and only one saturate raises bad cholesterol, so this weakens the correlation between saturated fat, bad cholesterol, and then heart disease, making that correlation nearly nonsensical.

Paradoxically, however, plant oils in most cases actually lower bad and then raise good cholesterol, thus triggering our bodies to remove cholesterol from our cells and possibly even from our plaques. Monounsaturates tend to raise our good relative to bad cholesterol. Polyunsaturates, both from the omega-6 and -3 categories, tend to raise our good cholesterol even more. In that sense, these oils do not affect the initial cause of heart disease one way or the other; however they prevent its later development, and possibly help in reversing it.[379,380,381]

Given all of this, we can rank dietary fats in their effects on cholesterol; at the top are the ones that are more prone to raise total cholesterol and possibly raise LDL compared to HDL.[382,383] When descending down the list, they are more inclined to lower total cholesterol and lower LDL compared to HDL:

- Palm kernel oil: extremely high in palmitate and low in monounsaturates and polyunsaturates

- Coconut oil: extremely high in myristic and lauric

- Butter: high in palmitate and also contains myristic, lauric and low amounts of monounsaturates and polyunsaturates

- Red Palm oil: high in palmitate but balanced in mono and low in polys

- Lard: higher in stearate but balanced with monounsaturates and polyunsaturates

- Tallow: higher in stearate

- Duck and chicken and some rodent fats: higher in monounsaturates and polyunsaturates, relative to other animal fats

- Marine fats: equal parts saturates, monounsaturates and polys

- Olive and Avocado oil: high in monos

- Peanut, almond, cashew, pecan oil: high in monos, medium in polys

- Various seed oils: corn, canola, safflower, cottonseed, soybean, etc., various ratios of monounsaturates to polyunsaturates with ranges of omega-6 to -3

SCIENTIFIC STUDIES

Cholesterol and atherosclerosis

Finally, scientists have conducted numerous studies on the lipid hypothesis over the decades—but the results are mixed and confusing. Furthermore, many organizations have strong and vested interests to get certain results, which means there is ample room for manipulation in either direction. Generally, it appears that one part of the hypothesis is supported by evidence—that is, higher total cholesterol, and higher LDL compared to HDL, in the blood correlates with a greater risk for atherosclerosis. However, the other part of the hypothesis is not supported by evidence—that is, that higher levels of animal or saturated fat of any kind increase total cholesterol, and LDL relative to HDL, and thus contribute to atherosclerosis. In other words, if you eat more animal fats or saturated fats, you do not increase your risk for atherosclerosis.

Higher total and bad cholesterol in the blood is correlated with heart disease. When some animals contain high amounts of cholesterol in their blood, both good and bad, they are more likely to develop atherosclerosis. When born with genetic aberrations for high cholesterol, animals and humans tend to develop atherosclerosis. When born with the same for low cholesterol, they avoid atherosclerosis. In humans, low levels of cholesterol in the blood, and even extremely low levels, protect against atherosclerosis. And when born with presets for lower amounts of bad cholesterol, humans live longer. And human infants have extremely low levels of cholesterol (even though human milk is 40% saturated, mostly from palmitate). When studying various groups in the US, scientists additionally find correlations between high levels of total and bad cholesterol and atherosclerosis —in people both without heart disease and preexisting heart disease.[384]

Other studies show that this same rule applies to different countries and distinct populations—and that when people migrate from areas with low cholesterol to live in other areas with high cholesterol, they in turn then raise their cholesterol—and all of this results in more heart disease. Conversely, in regions where people maintain low levels of cholesterol, they do not develop heart disease.[385]

Other studies show that when LDL cholesterol is under 100mg/dL in whole populations, they are at lower risk for heart disease — but that risk increases as that number increases from there. These studies also show that high total cholesterol when young correlates with heart disease later in life. When reducing total cholesterol by 1%, humans lower their probability of heart disease by 2%. At age forty, if we lower our total cholesterol by 10%, we then reduce our likelihood of CHD by 50% but that drops to 20% at age 70. Other studies show that using statins reduces cholesterol and LDL and heart disease by 20%.[386] One meta-analysis showed that with higher total cholesterol, people exponentially increased risk for heart disease and all-cause mortality.[387,388]

Another study showed that higher levels of IDL, VLDL, and LDL, but not total cholesterol, are correlated with greater potential for heart disease.[389] More confusingly, another study showed that even when total cholesterol is normal, yet consists of small LDL and HDL particles, our potential for heart disease is increased.

(HDL and LDL come in different particle sizes; and when smaller, HDL is even more damaging, and LDL is even less effective in "reverse cholesterol transport.").[390] The Framington Heart Study shows that when total cholesterol increased, overall mortality rose by 5% over thirty years and heart disease by 9%.[391] Many other studies suggest or support these same patterns—that is, that total cholesterol and bad cholesterol and even small particles of cholesterol are correlated with osteosclerosis. But keep in mind that these correlations are not all that strong. Furthermore, these correlations do not mean that animal fats, or saturated fats, are causing these increases in cholesterol and thus atherosclerosis. As noted many other factors also influence cholesterol.

Animal or Saturated Fat and Atherosclerosis

Overall, scientific studies do not support the correlation between saturated fat and cholesterol and heart disease—even though, paradoxically, some of these fats tend to raise total and bad cholesterol if only to some small degree. One meta-analysis of both randomized trials and observations studies did not show any correlation between reducing the intake of saturated fat and then reducing CVD or total mortality; in fact, saturated fat was protective for stroke. They say this is due to saturated fat not increasing small lipoproteins, which are more correlated with heart disease than large lipoproteins. They add that whole fat dairy, unprocessed meats, and dark chocolate do not increase the risk for CVD.[392] Even the American Heart Association, through analyzing various meta-analyses, concluded they could not find any association between reducing saturated fat and increasing polyunsaturated fat, and then reducing CVD.[393] Another meta-analysis of twenty-one studies showed that increasing intake of saturated fat does not result in increased risk for CVD, CHD, or stroke.[394] Another study showed mixed results.[395]

Additionally, scientists generally cannot find associations between specific fats, like butter and palm oil, with heart disease. Even while butter is at the top of the list for increasing both total and bad cholesterol, one meta-analysis showed either an insignificant or neutral association between butter and heart disease, mortality and diabetes. Another study shows that red palm oil, also at the top of the list, not only does not increase but actually decreases the probability of heart disease while also, paradoxically, reducing total cholesterol, atherosclerosis, and blood pressure.[396]

We also know from ancestral consumption that many, and even most humans, evolved to consume either moderate or enormous amounts of animal cholesterol and fats from various animals, including their brains, eyes, eggs, body fat and dairy. And humans do not evolve to consume foods that reduce our potential for survival—or that kill us. We do not have any reason to believe that ancient foragers suffered from heart disease; in fact, we mostly have reasons to believe that they never developed this or any other disease, even when living into their sixties. Additionally, more recent foragers and herders do not show any signs of heart disease, even when consuming large amounts of animal fats from reindeer and buffalo. The Plains Indians, who consumed enormous amounts of bison fat, did not show any signs of heart disease.[397] The Sami of Scandinavia and the Nenets of Siberia, who herd reindeer, eat enormous amounts of animal fat, including the brain and eye, but not much dairy (as reindeer do not produce much dairy), and do not develop heart disease.[398] The Inuit, who eat mostly marine fat combined with reindeer fat, do not have heart disease.[399] More or less, all traditional herding cultures in Africa, Mongolia, Europe and all across the world, do not develop heart disease even though consuming on average about the equivalent of eight to ten tablespoons of ruminant body and dairy fat per day—both of which are high in palmitate and moderate to low in monounsaturates and really low in polyunsaturates. Also, many of these cultures do not have access to fat from plants or smaller animals, which tend to contain more monos and polys. Cultures across the world that consume high amounts of red palm oil do not develop heart disease. And cultures, mostly in Polynesia, that consume high amounts of coconut oil, do not show any signs of heart disease.

SUMMARY

So scientists can trace correlations between increased total and bad cholesterol in the blood and atherosclerosis; however, those correlations are not necessarily all that strong and appear to express themselves over many decades. Additionally, many factors contribute to increased cholesterol, including genetics, high calories, high carbs, high sedentism as well as many others, both known and possibly unknown.

However, scientists cannot trace reliable correlations between animal fat (or saturates or specific saturates) and atherosclerosis or even anything bad. Furthermore, nutritional physiology does not support the correlation—as cholesterol and saturated fats do not cause the initial damage to our arteries. Ancestral consumption does not support the correlation—as the primary, or one of the most primary foods, for our ancestors was copious amounts of animal fat. In any case, and somewhat paradoxically, you can increase your amount of monounsaturates and polyunsaturates and possibly reduce your chances of atherosclerosis. But at the same time, our ancestors probably never consumed fats in these ratios—so this is experimental with unknown consequences.

CHAPTER 27

PREMISE 6.4 – OMEGA-3 AND OMEGA-6

NUTRITIONAL PHYSIOLOGY

As we know, the polyunsaturated fatty acids are called "essential" because we cannot make them, so we must eat them in our diet. Nature makes about eleven polyunsaturates but we only consume five of them in higher amounts. They are all long-chain fatty acids, but some of them have shorter chains and others have longer chains. Generally, the shorter chains are found in plants and animals while the longer chains are found only in animals. Polyunsaturates are either omega-6 or omega-3 which refers, chemically, to where the double bond is located on the carbon chain, either 6 or 3 chains down, but this also has implications for our health.

Only two of the omega-6s predominate in our diet, including the shorter chain, linoleic acid (LA), and the longer chain, arachidonic acid (AA). LA is more abundant in most seeds, including grains, nuts, and legumes so that animals that specialize on these foods, including humans, contain greater amounts of LA in their adipose—which they then convert into AA— which then predominates in their phospholipids, especially in their liver and brain. Humans in some cases cannot necessarily convert enough LA into AA, so they must receive their AA from other animals.

Only three of the omega-3s predominate in our diet: the shorter chains, alpha linolenic acid (ALA) and the longer chains, eicosapentaenoic acid (EPA) and docosahexaenoic acid (DHA). ALA is found in some select seeds, like flax and chia, but most abundantly in algae, lichens, leaves and grasses so that animals that eat these foods contain greater amounts of omega-3, including fish, seals and ruminants like bison, reindeer, cows and sheep. In most cases these animals store ALA in their adipose and then convert that into EPA and DHA, which they then use in their phospholipids, especially in their brain and eyes. Most animals, including humans, can convert ALA into EPA and DHA, but not necessarily enough.

Usually our muscles prefer omega-6s and our brain more omega-3s if both of these are ample enough in our diet. But if one of the omegas is missing or imbalanced with the other, then our phospholipids potentially utilize them in ways that are detrimental. In particular, when humans do not consume enough omega-3, they use too much omega-6 in their brain — which then causes inflammation, impairment and possibly depression, anxiety and other forms of mental illness.[400,401,402]

At times our cells pull the polyunsaturates from their membranes and convert them into the hormones called eicosanoids that operate inside our cells. If our membranes contain more of the omega-6, AA, we make more of certain kinds of eicosanoids, and if they contain more of the omega-3's, EPA and DHA, they make more of other kinds of eicosanoids — which then affect most systems throughout our body, including our immune and cardiovascular function. So for this reason, we need to keep our consumption of these fatty acids in balance with each other to avoid many detriments. We catabolize the omegas the same as any other fatty acid. Additionally, we only store the shorter chains in our adipose, mostly as LA, at about 15 to 20% of total fat but, when consuming more omega-3s, we store more of them.

If under-consuming polyunsaturates, we encounter all sorts of detriments because they are "essential." When under-consuming the omega-6s, mostly in the form of linolenic acid, we typically suffer skin problems, impaired growth and development, increased inflammation, weakened immunity, and hormonal imbalances.[403] When under-consuming omega-3, we suffer cardiovascular issues, cognitive impairment, vision problems, inflammatory conditions, mood disorders, impaired growth and development.[404] When consuming the omegas in imbalanced ratios to each other, we suffer other detriments as well. When consuming too much omega-3 to -6, we suffer increased inflammation, cardiovascular issues, and dysregulation in our immunity, brain and signaling, while also trending us towards the diseases of civilization.[405] When consuming too much omega-6 to -3 in the wrong ratios, we suffer excessive inflammation, chronic diseases (cardiovascular disease, autoimmune, cancer and the diseases of civilization, psychological disorders, and other conditions like arthritis, inflammatory bowel disease).[406]

ANCESTRAL CONSUMPTION

Our theoretical ancestor consumed less fat but about 30 to 50% of that fat as polyunsaturates from such sources as leaves, insects, and nuts in ratios of about two to four parts omega-6 to one part omega-3. But human foragers consumed more fat, and more animal fat, which lowered their consumption of the polyunsaturates. In warmer environments, they tended to consume more polyunsaturates and more of them as omega omega-6 because they ate more seeds and animals that ate those seeds. When inhabiting colder or marine environments, our ancestors consumed more omega-3 to -6 because they ate mostly animals that consumed grasses, lichens, and algae. Some scientists claim that most foragers consume about two parts omega-6 to one part omega-3; however, I cannot find much evidence to support this claim.

Agriculturalists consumed more polyunsaturarates than foragers because they consumed more seeds. While most of these seeds were high in omega-6, they also selectively cultivated seeds higher in omega-3, such as flax, chia and poppy while also eating animals pastured on grass. So on average I estimate agriculturalists consumed about four to ten parts omega-6 to one part omega-3. Herders consumed enormous amounts of fat from ruminant tissues and dairy—which are both low in polyunsaturates, at around 3 to 5% of total fat with higher amounts of omega-3. The ratios in reindeer are about one to one; sheep, about two to one, and bison and cattle, about four to one. To receive longer chain polyunsaturates, including AA, DHA and EPA, they consumed the brains, eyes and other organs from their herds.

During the modern we started to consume excesses of polyunsaturates from domesticate seeds like canola, corn, cottonseed and soybean, and turn them into plant oils. In some cases these oils have somewhat balanced ratios of omega-6 to -3, like canola, but in most cases, they have imbalanced ratios, like cottonseed and corn, that contain enormous amounts of LA but little or no ALA. At the same time, because we fed our livestock on seeds instead of grass, they contained higher amounts of LA as well. So during the modern we potentially consumed about two to three times as many polyunsaturates and imbalanced ratios of omega-6 to -3, trending us towards the diseases of civilization.

ESTABLISHMENT RECOMMENDATIONS

The establishment recommends that we consume about 14 to 30% of our fat from polyunsaturates—or about 11 to 22 g of polyunsaturates per day as optimum while setting adequate amounts for LA at around 11 to 15 g; and adequate amounts for the omega-3, AA, at about 1.3 g for men and women. Some parts of the establishment recommend about 0.5 g of the longer chain omega-3s, EPA and DHA per day. The establishment does not recommend any ratio between omega-6 and -3 but, nonetheless, from the data above, they inadvertently recommend ratios at around ten to twenty parts omega-6 to one part omega-3—even though considerable evidence suggests this ratio is problematic, causing chronic inflammation, heart disease, and auto-immune disorders. In some cases, they do not specify shorter or longer chains of these fats—even though some people cannot necessarily convert enough of the shorter into longer chains. As noted before, the nutritional establishment recommends more polyunsaturates overall because they reduce bad cholesterol and raise good, and presumably, help prevent heart disease.

SCIENTIFIC STUDIES

All scientists acknowledge that the polyunsaturates, both the omega-6s and -3s, are essential to our health. But they vary on the amounts we should consume and the ratio of omega-6 to omega-3, with many studies not supporting the establishment recommendations. On the upside many studies show that the omega-6, LA, at the amounts recommended by the establishment, helps reduce heart disease, metabolic syndrome, liver fat, and type 2 diabetes while improving insulin sensitivity and glycemic control. These studies also confirm that these amounts do not increase inflammation, oxidation or obesity in the body.[407,408]

However, other studies suggest caution about the consumption of polyunsaturates or specifically, LA at the amounts recommended by the establishment. As noted before, polyunsaturates are more subject to oxidation, due to their chemical structure, changing into numerous compounds, called OXLAMS, some considered detrimental, others beneficial. However, in total, these compounds are potentially more detrimental as they lead to more oxidative stress and platelet aggregation, chronic inflammation, cardiovascular disease, and brain problems. Paradoxically, while LA reduces bad cholesterol and raises good, OXLAMS nonetheless are implicated in oxidizing cholesterol more—which, if you remember, is one of the steps leading to heart disease. At 20 g per day, LA did indeed increase oxidative stress and endothelial malfunction — which could contribute to heart disease.[409]

Furthermore, studies show that reducing LA in the diet then reduces the four OXLAMS considered harmful.[410] Many studies do not show any correlation between LA and improvements in our health in any way, and otherwise suggest they promote inflammation and cancer.[411] Additionally, the longer chain polyunsaturates, AA, which are essential to our membranes and eicosanoids, are also considered to increase inflammation in excess. And while LA is converted into AA, we, paradoxically, do not increase AA by consuming more LA.[412, 413, 414]

RATIOS OF OMEGA-6 TO OMEGA-3

Many studies show that, when balancing our ratios of omega-6 to -3, humans see reductions in the incidence of various diseases and improvements in their overall health, at somewhere around two to six parts omega-6 to one part omega-3.[415,416] When people already showed signs of cardiovascular disease, or even experienced episodes like heart attacks, they showed 70% prevention in cardiovascular disease when consuming four to one ratios, as well as reductions in the overall diseases of civilization.[417] Ratios of about three to one reduced colonic cancer, whereas, oddly, the ratio of four to one did not have any effect. Closer ratios also improved breast cancer; ratios of three to one suppressed inflammation for rheumatoid arthritis. Five to one was beneficial for asthma, whereas ten to one was detrimental. Their balance is also important for brain development.[418]

More distant ratios are associated with increases in many diseases, including cardiovascular disease, cancer, osteoporosis, and inflammatory and autoimmune diseases, whereas increased levels of omega-3 polyunsaturated fatty acids (PUFAs) (a lower omega-6/omega-3 ratio) exert suppressive effects. While closer ratios help in the prevention of disease, scientists of course have not yet determined how much polyunsaturated and ratios of omega are optimal —

as what optimizes disease prevention does not necessarily optimize our metabolism.[419]

SUMMARY

Nutritional Physiology

When making recommendations for fats, I think more along the lines of the preponderance of the evidence—rather than the absoluteness of the truth. As already noted, nutritional physiology determined that we should consume so many of our calories from carbs and the remainder from fats. And we humans need somewhere around 180 carbs per day, perhaps 50 or so more, and then more for strenuous exercise.

When applied to our theoretical consumer, we yield these results:

CALORIES 2300.
PROTEIN: 15% (86 g)

TOO LITTLE FAT AND TOO MUCH CARBS

Fats: 35% (89 g) Carbs: 50% (286 g)

BALANCED FAT AND CARBS:

Fats: 40% (102 g) Carbs: 45% (258 g)
Fats: 45% (115 g) Carbs: 40% (230 g)
Fats: 50% (127 g) Carbs: 35% (200 g)

TOO MUCH FAT AND TOO LITTLE CARBS

Fats: 55% (140 g) Carbs: 30% (170 g)

In regards the ratio of fatty acids, humans must acquire the essential fatty acids in their diet, the polyunsaturates from both the omega-6 and -3 in certain amounts and ratios—or otherwise trend towards detriments or disease. Overall we probably need at least 5 to 10 g of polyunsaturates per day, but more may possibly provide benefits—even as over 15 to 20 g per day may trend us towards detriments. Also we want these omegas in about two to eight parts omega-6 to one part omega-3 and probably want to make sure some of them come from the longer chains found in animal fats—that is, AA, EPA and DHA because we cannot make these from the shorter chains.

First, let's consider that our phospholipids use fatty acids in the following approximate ratios—even though we only use small and even tiny percentages of our dietary fats for these membranes.

Saturates – 35%
Monounsaturates – 30%
Polyunsaturates – 35% (mix of omega-6 and -3)
*Ranges somewhat influenced by diet

Second, let's consider that our adipose stores fatty acids in these approximate ratios:

- Saturates: 20 to 35%

- Monounsaturates: 45 to 60%

- Polyunsaturates: 5 to 15% (higher amounts of omega-6 to -3)

Third, for catabolism, we presumably can use any fatty acids available as our cells do not appear to prefer one type over another: they all generate the same amount of ATP per mass. At the same time, we can convert or make almost all fatty acids in our body, except for the polyunsaturates.

So how do we draw conclusions from this data? I could argue that, provided humans receive enough of the polyunsaturates in their diet, they can eat almost any type of fat or ratio of fatty acids and survive and even thrive. But I lean more towards consuming fats somewhere close to the ratios where we store them in our membranes, and most of all in our adipose. So in short, nutritional physiology suggests that, first of all, we should make sure we consume enough polyunsaturates in the proper ratios and otherwise consume our fatty acids in the ratios that we store them, leading to these results:

- Saturates: 30 to 40%

- Monounsaturates: 40 to 50%

- Polyunsaturates: 20% with about four to eight parts omega-6 to one part omega-3.

Ancestral consumption

Most of our human ancestors consumed this amount of fat in their diet as applied to our theoretical consumer at 2300 calories:

- Fats: 30% (77 g) Carbs: 55% (316 g)

- Fats: 35% (89 g) Carbs: 50% (288 g)

- Fats: 40% (102 g) Carbs 45% (258 g)

- Fats: 45% (115 g) Carbs 40% (230 g)

- Fats: 50% (127 g) Carbs: 35% (200 g)

- Fats: 55% (140 g) Carbs: 30% (170 g)

- Fats: 60% (153 g) Carbs: 25% (144 g)

- Fats 70% (179 g) Carbs: 20% (115 g)

- Fats 80% (205 g) Carbs: 15% (86 g) (When consuming carbs in these amounts, humans usually consume more protein for gluconeogenesis and ketogenesis.)

With these ranges in the ratios of fatty acids:
- Saturates: 30 to 50%

- Monounsaturates: 40 to 60%

- Polyunsaturates: 10 to 30% (with three to eight parts omega-6 to one part omega-3)

On average, I think most humans across all eras consumed their fats about like this:
- Fats: 40% (102 g). Carbs 45% (258 g)

- About 30% saturated.

- About 50% monounsaturated.

- 20% polyunsaturated with six parts omega-6 and one part omega-3, and combinations of shorter and longer chains.

Establishment Recommendations

The establishment recommends these ranges of fat as applied to our theoretical ancestor:
2300 calories
- 15% protein (86 g)

- Fats: 20% (51 g) Carbs 65% (374 g)

- Fats: 25% (64 g) Carbs 60% (345 g)

- Fats: 30% (77 g) Carbs 55% (316 g)

- Fats: 35% (89 g) Carbs 50% (286 g)

They then recommend that we consume:
- Saturates: less than 20%

- Monounsaturates: about 60%

- Polyunsaturates: about 20% at about thirteen parts omega-6 to one part omega-3 as the shorter chains.

In this case, establishment recommendations are not in accord with nutritional physiology, ancestral consumption, or scientific studies, in the amount of fat because, relative to the others, the establishment recommends low amounts of fat and high amounts of carbs. The establishment also differs on the percentages of the various fats as they accentuate low amounts of saturates and high amounts of monounsaturates and even polyunsaturates. But our ancestors never or rarely consumed fats in those percentages — as they do not exist in our evolutionary diet. The establishment also recommends high amounts of polyunsaturates as LA, as well as high amounts of omega-6 relative to omega-3 — but these recommendations do not accord with nutritional physiology, ancestral consumption or scientific studies.

In recommending fats in these percentages, the establishment is following their beliefs about the Lipid Hypothesis in three ways. First, they are affected by the idea that total fat in the diet contributes to heart disease — so they recommend lower amounts of fats and exorbitant amounts of carbs. Second, they still believe the idea that saturated fat causes heart disease so they lower the amount of saturates. Third, they believe that high amounts of polyunsaturates protect against heart disease, so they raise the amounts of polyunsaturates. But as noted the lipid hypothesis is not really proven, especially because animal fats do not cause heart disease. If anything, we have more reason to believe that animal fats are protective against heart disease since so many of our ancestors and contemporaries eat diets way high in animal and saturated fats without any signs of heart disease — or the diseases of civilization.

RECOMMENDATIONS

So I am recommending that we disregard establishment recommendations on the amount of fat, the percentages of fats, and the amount and ratios of omega-6 to omega-3, because of their lack of accord with all other methodologies. Instead, we should focus on nutritional physiology, ancestral consumption, and scientific studies to draw our conclusions. So I recommend that we consume our amounts of fats in these ranges as applied to our theoretical consumer:

- Calories: 2300

- Assuming 15% protein

- Fats: 35% (89 g) Carbs: 50% (288 g)

- Fats: 40% (102 g) Carbs: 45% (258 g)

- Fats: 45% (115 g) Carbs: 40% (230 g)

- Fats: 50% (127 g) Carbs: 35% (200 g)

With these ranges in the ratios of fats:

- Saturates: 30 to 40%

- Monounsaturates: 40 to 50%

- Polyunsaturates: 20% (around four to eight parts omega-6, to one part omega-3)

However, I recommend this more precise amount for our theoretical consumer, given his level of fitness and exercise:

Calories: 2300
Protein as a percent of calories: 15

- Carbs: 45% of calories (260 g) (ten apples or twenty slices of whole wheat bread)

- Fats: 40% of calories (100 g) (seven tablespoons of pure fat or oil)

- Saturates: 30% (30 g)

- Mononunsaturates: 50% (50 g)

- Polyunsaturates: 20% (20 g) at four to eight parts omega-6 to one part omega-3

When exercising more at moderate levels, you need more fat in your diet and about the same amount of carbs. When at more intense levels, you need more carbs but also fats because your basal metabolism increases. To reduce or maintain weight, you should consume around 150 to 200 carbs per day as your brain and other tissues catabolize that amount—and thus those carbs are not then stored as fat. At the same time, you probably want to reduce your fat while eating enough of the polyunsaturates—even while knowing, paradoxically, that eating more fat, within reason, does not appear to make you as fat as overeating carbs.

Generally, this accords with the preponderance of the evidence pertaining to nutritional physiology, ancestral consumption and scientific studies. Interestingly, this also accords, somewhat anyway, with human milk:

- Protein: 6%

- Carbs: 40%

- Fats: 52%:

- Saturated: 40%

- Monounsaturated: 35%

- Polyunsaturated: 25% (mostly omega-6 but dependent upon diet)

When eating this amount of fat, we provide our body with enough ATP to optimize our ATP and our energy. We also provide all of our tissues that prefer fats with enough fats; and the same for our tissues that prefer carbs. When consuming our fats in these ratios, we easily provide enough of all fatty acids for our phospholipids, including the omega-6s and -3s, especially when eating the longer chain versions of these fats. We provide the right balance of omega-6 to -3 to balance our eicosanoids. We also consume fats in the approximate ratios as we hold and utilize them inside our bodies. Interestingly enough, even if we believed in the lipid hypothesis, these ratios would actually lower bad cholesterol and raise good cholesterol, although only slightly.

To consume fats in these amounts and ratios, we only need to eat like our ancestors—that is, mostly consume animal fats from meat, tallow, dairy and eggs while also consuming plant fats from olives and avocados, as well as from specific seeds containing both LA and AA. Otherwise you should specifically consume eggs and seafood at least several times per week to acquire your longer chain polyunsaturates, AA, EPA and DHA.

CHAPTER 28

PREMISE 6.5 – MACROMINERALS

ELECTROLYTES

Nutritional physiology

We humans utilize six different electrolytes: potassium, magnesium, phosphorous, calcium, sodium, and chloride. These electrolytes control the flow of liquids in our bodies, especially from our blood into our cells and back again while carrying nutrients and wastes along with them. They also control the flow of electricity, causing our cells to relax or activate at various levels of intensity, which in turn causes our muscles to relax or constrict and our neurons to relax or fire.[420] In some senses, the macronutrients create the structure of our cells, the membranes, the organelles as well as the ATP, but the electrolytes actually animate our cells, to move, to think and to feel. As such they are extremely important to our metabolism.

Unlike most nutrients, macrominerals are atoms, not molecules, so they are not broken down into smaller units which means we can recycle and reuse them endlessly. In most cases, we regulate the amount of these minerals in our blood and other tissues, keeping them within normal ranges, by taking up only certain amounts from our gut and then retaining or expelling certain amounts from our kidneys. We absorb 90% of our potassium from our guts into our blood; 30 to 40% of our magnesium; 60 to 80% of our phosphorous; and 90% of our sodium and chloride.[421,422,423] Once these minerals flow through our kidneys, we either recirculate them back into our blood if our body needs them or excrete them in our urine. Additionally, our body has multiple mechanisms for determining if we need more or less of them. However, we regulate calcium more closely, only absorbing about 20 to 40% because this mineral is potentially damaging to our body and blood.[424] Additionally, our bodies constantly break down old bone, recycle the calcium, and then rebuild new bone. Scientists do not know why we absorb so many electrolytes only to excrete most of them.

Despite our ability to absorb and retain electrolytes, we can still suffer deficiencies which result in fatigue and weakness as well as muscular, neurological, cardiovascular and fluid problems; however, these conditions are rare, at least in most cases. We can also suffer from excesses of electrolytes, with too much potassium (hyperkalemia), too much magnesium (hypermagnesemia), etc., but this only happens with heavy supplementation or bodily dysfunctions like failing kidneys—at least in most cases.[425]

But modern humans regularly over-consume salt, or sodium and chloride, and when that happens, we retain the salt and increase the volume of our extracellular fluid, which pulls water from our cells, trending us towards high blood pressure which then damages our arteries and trends us towards heart disease. We possibly can over-consume calcium from dairy, which in some cases blocks the absorption of magnesium, leading to deficiencies, especially when supplementing with too much vitamin D and too little vitamin K, which leads to calcium spiking in our blood. If we traumatize parts of our body, we are prone to calcifying those tissues.[426]

Ancestral consumption

I break down ancestral consumption of electrolytes into four categories: 1) Omnivorous, foraging primates; 2) Carnivorous, foraging humans (who ate some plants but not much); 3) Agricultural humans; 4) Modern humans with the SAD diet. The first category, omnivorous, foraging primates, includes monkeys, apes, hominoids and humans (especially the humans who consumed lots of plants). Generally, all these primates consumed lots of fruits, tubers, seeds and leaves as well as animals. Since all these foods collectively contain similar amounts of electrolytes, these primates thus ate similar amounts of electrolytes per day, including enormous amounts of potassium, but moderate amounts of magnesium and phosphorous and lower amounts of calcium and salt (sodium chloride). They consumed small amounts of salt from animal blood, mineral deposits, salt water and aquatic plants. They also consumed smaller amounts of calcium because both plant and animal tissues contain little calcium. For example, kale and almonds are the richest forms of calcium in our diet outside of dairy and sardines. But we need to consume ten cups of kale and three cups of almonds to equal just one cup of milk. Human foragers occasionally eat the bones of baby birds and small fish but, from what I can tell, this is infrequent and otherwise they do not seem to select for mineral deposits or special foods high in calcium. In other words humans only consume enormous amounts of calcium from dairy. But overall these primates consumed enormous amounts of electrolytes and then peed them out.[427]

The second category includes carnivorous, foraging humans, such as the Inuit and Plains Indians, who consumed fewer electrolytes than all humans. From the tissues and blood of their prey, they consumed only about one fourth the amount of electrolytes as compared ominvores. So they absorbed most of their minerals and then probably retained most of them—thus preventing deficiencies.[428] Meanwhile herders consumed mostly or only animal foods, but from dairy—which is extremely high in calcium, phosphorous, potassium while containing some magnesium, sodium and chloride.

The third category is agricultural humans who consumed mostly seeds, along with animal foods, including dairy and eggs, and smaller amounts of fruits. As compared to omnivorous foragers, they consumed less fruit and more seeds, resulting in less potassium and more magnesium and phosphorus. They also consumed enormous amounts of dairy as well as salt used to preserve and flavor their foods. Overall they consumed higher amounts of nearly all electrolytes, as compared to their omnivorous foragers, with the possible exception of potassium.

The fourth category is modern humans with the SAD diet — the standard American diet. These diets consist of many of the same foods as agriculturalists, except moderns refine their seeds, stripping away much of the magnesium and phosphorous. And we refine other plants to make sugar while stripping away all electrolytes. At the same time, we consume high amounts of dairy, meat and salt. In the end moderns consume lower amounts of potassium and magnesium and higher amounts of phosphorous, calcium and sodium/chloride. In this particular case, we are experiencing detriments from the amounts of electrolytes and their ratios to each other: too much salt relative to potassium; too much calcium relative to magnesium; and possibly too much phosphorous — all of which probably trend us towards the diseases of civilization.[429]

Establishment recommendations

Generally, as noted above, the establishment recommends that we consume about 5 g of potassium, 1 gram of phosphorous, 310 millig of magnesium, 1 gram of calcium and 1.5 g of sodium/chloride combined.

Catagory	Potassi-um	Phospho-rous	Magne-sium	Calci-um	Sodium/Chlorine
Foraging, omni-virous primates, including humans	5 to 8 g	1 gram	500 mg	400 mg	500 mg
Foraging, carni-virous primates, including humans	1.5 g	1 gram	150 mg	100 to 500 mg	400 mg
Agricultural humans	4 g	2 g	400 mg	1 gram	1.5 g
Modern Humans	2 to 4 g	2 g	100 mg	1 gram	2 g
Establishment recommendations	3 g	3 g	310 mg	1 gram	1.5 g
Recommendations	5 g	1 gram	300 mg	500 g	1 g

TABLE 28.1- AMOUNT OF ELECTROLYTES PER GROUP OF HUMANS.

Summary and recommendations

So humans are adapted to consume various amounts and ratios of electrolytes—at least to some extent—due in part to our ability to selectively uptake and then retain or expel them—and indefinitely, given that they are all atoms and thus never broken down into smaller particles. We also adapt to varying levels of electrolytes through natural selection, at least in some ways. Europeans, for example, evolved to tolerate higher amounts of salt as compared to Africans, who thus are more prone to develop high blood pressure from consuming salt.[430] I surmise that we are possibly adapted to greater or lesser amounts of calcium and phosphorous from dairy. More carnivorous primates, including humans, have possibly evolved for fewer electrolytes overall.

But the question remains: even though we can survive and reproduce on various levels and ratios of electrolytes, do we optimize our metabolism in some way by eating them in more specific ways? I cannot fully answer this question, but offer some general direction. Generally we probably cannot under-consume electrolytes in our modern diet as supported by nutritional physiology, ancestral consumption and even scientific studies.

If some of our ancestors did not under-consume electrolytes from eating just animal tissues, we probably cannot from eating our mixed and modern foods—even though many of them are stripped of electrolytes. We can just absorb and then retain more of them.

Even though this is the case, the establishment does recommend high amounts of them, including calcium, phosphorous and salt as compared to all of our ancestors prior to agriculture. So in short, if your diet is balanced, you should never suffer any deficiencies—except maybe temporarily in cases of extreme exertion and heat.

All four methodologies confirm that we cannot over-consume electrolytes from regular food. However we can over-consume electrolytes from too many supplements, and too much salt and dairy. So to avoid over-consuming electrolytes, you should not consume supplements and use salt and dairy in moderation.

We can consume electrolytes in imbalanced ratios to each other—and generate problems for ourselves—including too much salt relative to potassium, which can increase hypertension, stroke, heart disease, kidney disease, muscle cramps, and nerve issues.[431,432,433,434] When consuming too much calcium relative to magnesium, we lower the uptake of magnesium, leading to deficiencies—which in turn trends us towards muscle cramps, bone issues, cardiovascular problems and kidney stones.[435,436] This can happen by consuming too much dairy and not enough seeds, or eating refined seeds with most of their magnesium removed, or eating whole seeds that are not properly processed to make their magnesium bioavailable.

If you eat the diet recommended in this book, while avoiding supplements and excesses of salt and dairy, you naturally consume enough electrolytes and balance them as well—even though you are probably peeing most of them away. In this case you are consuming electrolytes in similar ranges as omnivorous primates, and somewhat similar to establishment recommendations, at about these amounts: potassium: 5 g; phosphorous: 1 g; magnesium: 300 mg; calcium about 750 mg; and sodium and chloride: 1 g.

CHAPTER 29

PREMISE 6.6 - PHYTONUTRIENTS

I also recommend that we consume a diversity of phytonutrients. Phytonutrients are generally defined as "secondary plant compounds" which, as mentioned before, are compounds used by plants to defend against predators, including bacteria, fungi, insects and animals and even humans. While possibly providing detriments to humans, to some slight degree, these compounds provide much greater benefit if consumed in moderation. They are also common in our diet and appeal to our sense of sight.

For tens of millions of years, our ancestors consumed primarily fruits and leaves rich in phytonutrients—and during that time, evolved to mostly neutralize their detriments and even turn them into benefits. However, we humans are not necessarily dependent upon them for our health because many of our ancestors, such as the Inuit and other arctic people, did not consume many or even any of them. But whenever they were available, our ancestors tended to consume them: the Plains Indians added berries and other plant foods to their pemmican and the Inuit, in their short summer months, added berries, leaves and lichens to their diet even though they did not provide enough macronutrients to matter.

These phytonutrients include polyphenols, terpenoids, resveratrol, flavonoids, isoflavonoids, carotenoids, limonoids, glucosinolates, phytoestrogens, phytosterols, and anthocyanins. In some cases they serve the same purpose in us as they do in plants—that is, they boost our immune systems and protect us from microorganisms.[437] Additionally, many phytonutrients are associated with many of these benefits: anti-microbial, anti-oxidants, anti-inflammatory, antiallergic, anti-spasmodic, anti-cancer, anti-aging, hepatoprotective, hypolipidemic, neuroprotective, hypotensive, diabetes, osteoporosis, CNS stimulant, analgesic, protection from UVB-induced carcinogenesis, immuno-modulator, and carminative.[438] In some cases, these benefits are proven or somewhat so—but in other cases suggested.

Plant pigments

One category of phytonutrients is plant pigments—which give plants their colors, including green, yellow, orange, red, purple, etc. Plants use these pigments to absorb certain wavelengths of light used in photosynthesis and to protect themselves from damaging wavelengths—and for the other purposes mentioned. These plant pigments include: carotenoids; anthocyanins; and betaines.

Carotenoids are broken down into two categories—the "provitamins" that are precursors to retinal, otherwise known as vitamin A, and the non-provitamins. The provitamins include beta carotene, lycopene, lutein, and zeaxanthin. Reddish and orangish in color, beta carotene generates the most Vitamin A and is found in carrots, greens and many other plants. Reddish in color, lycopene generates trace amounts of Vitamin A and is found in tomatoes and other plants. Yellowish and reddish in color, lutein and zeaxanthin also generate trace amounts of vitamin A and are found in green, leafy vegetables like kale, spinach, broccoli, etc. These pigments are essential to our health because we store them in our eyes, as well as our brain, to protect our eyes from damaging light while helping us see better; they also improve cognition and provide other benefits to our body.[439] The other carotenoids, which are not provitamins, are anthocyanins (also considered flavonoids) and betalains, which are reddish, blueish and purplish in color.[440,441] Anthocyanins are associated with reducing blood pressure, heart disease, neurological diseases, and cancer growth. Betalains are associated with combating and improving inflammation, detoxification and slowing cancer. Generally, all carotenoids are antioxidants, helping to keep free radicals in balance; and most carotenoids infuse our skin, protecting us from sun damage while giving our skin a more radiant and attractive sheen.[442] Though not defined as phytonutrients, chlorophyll is also a plant pigment, green in color, that provides benefits to humans, including skin healing, cancer protection, and weight loss.[443]

Polyphenols and flavonoids

Another category of phytonutrients is polyphenols which are present in most, if not all, plant foods. Generally, polyphenols help improve lipid profiles, blood pressure, insulin resistance, systemic inflammation and diabetes, as well as cytotoxicity. One category of polyphenols is called flavonoids, and the specific ones found in Cocoa are associated with decreased risk of stroke and diabetes. Also, when fermented in the gut, polyphenols are converted into beneficial compounds with therapeutic effects.[444]

Organic acids

Organic acids include lactic acid, acetic acid, formic acid, citric acid, oxalic acid, uric acid, malic acid, and tartaric acid. Plants generate organic acids during photosynthesis and use them as intermediaries in the Krebs cycle and for other purposes. Microorganisms, including the ones in our colon, generate organic acids during fermentation.

To us organic acids taste sour or tart. We humans generally taste the nutrients our bodies most need—so presumably we need these organic acids even though they are not defined as nutrients. However, for whatever reason, scientists have not studied the function of these nutrients all that much—even as I think they are more important than we know. But we do know that we use some organic acids as intermediates in the Krebs cycle just like plants, suggesting they help optimize catabolism.

All of them tend to help with our digestion because they help lower the acidity of our stomach—which makes digestion more efficient. Many of them also operate as antioxidants, antibiotics and antivirals while helping to balance our PH throughout our bodies. Lactic acid strengthens our immune system, helps in digestion, helps our body absorb minerals and vitamins and protects against infections from microorganisms.[445]

Acetic acid is beneficial in many ways. We receive acetic acid from three sources: from fermented foods, like kombucha and kimchi; from the fermentation in our colon; and from alcohol which is converted into acetic acid. Nearly all of our cells then catabolize acetic acid into ATP through the usual pathways. Interestingly, our brains appear to prefer acetic acid, possibly over glucose, because this fuel improves our cognition as well as our resistance to depression.[446] Additionally acetic acid is beneficial in the catabolism of fatty acids and glucose while also improving insulin resistance, weight loss and cholesterol markers.[447]

Malic acid, found mostly in apples, combats fibromyalgia and improves dental, liver, and skin health.[448] Uric acid neutralizes free radicals in our blood. We humans generate uric acid from purines found in the DNA of the plants and animals we consume, most especially the ones high in protein (except for dairy). In balanced amounts uric acid is the strongest antioxidant in our blood but in excessive amounts, it is detrimental to our health, precipitating from our blood to form crystals and ultimately causing inflammation and gout.[449] Oxalic, formic, and tartaric acids are considered toxic in higher amounts and, fortunately, are in our foods only in trace or smaller amounts.

SUMMARY AND RECOMMENDATIONS

Certain phytonutrients provide certain benefits—so for that reason, you should consume many different kinds by eating many different types of plants. To receive the benefits of plant pigments, you should eat plants across all spectrums of colors—greens, yellows, oranges, reds, and purples—while focusing on the most relevant pigments, including beta carotene which is converted into vitamin A, as well as lutein and zeaxanthin which are used to protect our eyes and brain and improve cognition. You should consume any plants for polyphenols and flavonoids, including chocolate and coffee, if you like caffeine. And you should also consume different types of fruits and vegetables, as well as fermented foods, for their variety of organic acids. At the same time, you should consume plenty of soluble fiber, which results in the fermentation of phytonutrients that are then converted into smaller compounds that provide benefits.

CHAPTER 30

PREMISE 7 – TASTE AND SATIATION

I recommend that you make your food, or really your meals, taste delicious and satiating by including and balancing all five pleasurable flavors and "mouthfeels," including sweet, acidic, salty, umami, and fatty, while minimizing unpleasurable flavors and mouthfeels, including bitterness and astringency. We moderns are perhaps conditioned to think that nutritious foods should taste bad. But in reality they should taste delicious when designed well—and in fact, the more nutritious the meal, the better it should taste—according to the theory of evolution. Evolution designs us so that our pleasurable tastes direct us to nutritious foods and unpleasurable tastes direct us away from non-nutritious or toxic foods. So by this reasoning, we should use our sense of pleasurable taste to steer us towards the most nutritious foods.

Humans have five tastes and the four of them we consider pleasurable are correlated to nutrients we need. Sweet designates the presence of carbs and their underlying sugars. Acidic (sour and tart tastes) designates the presence of organic acids such as malic and citric acids, which I already addressed. Salty designates the presence of sodium, chloride, and potassium. As explained earlier, we evolved to taste salt, but not the other minerals, because sodium and chloride are rare in our environment so we need to taste them, to find them; we consume enough of all other minerals whether we taste them or not. Umami designates the presence of glutamate (one of the major amino acids in all proteins) as well as nucleic acids that are part of DNA, and as such this flavor directs us towards proteins like meats and beans, especially when those foods are properly processed. When eating these nutrients, we taste them and derive pleasure from them, which in turn drives us to continue to consume them. We also derive pleasure from "mouthfeels," that is, the way foods feel on our tongue, with fats providing pleasure in this way along with other sensations on our tongue. Meanwhile we also have unpleasant tastes and mouthfeels that drive us away from certain foods. The flavor, bitter, and the mouthfeel, astringency, designate the presence of anti-nutrients and toxins in foods and thus drive us away from them.

Interestingly we humans do not taste two of the most predominant foods in our diet—that is, starches and some forms of meat. We taste sweet in the sugars but when they are part of starches, we do not taste them so that boiled wheat, potatoes and quinoa taste bland or worse. But why? Scientists do not know the answer to this question but I theorize that, by the time we started to specialize on starch, our ancestors were already quite encephalized so they just knew, cognitively, that this food was nutritious and increased satiation afterwards.

We also oddly do not taste much of anything in some meats: raw or cooked chicken breasts for example. Even though these meats are packed with glutamate and nucleic acids, we do not taste them, even when cooked, because these molecules are not exposed.

So again, why would we evolve to eat foods that we do not taste? Probably because our ancestors evolved to eat these foods as part of the whole animal with the blood, guts, and organs intact—which are richer in flavor. Also, while our ancestors evolved to eat more meat, they certainly cooked their meat; however, they probably used other forms of processing more, including drying, smoking and fermenting because these preserved the meat for later use. And these types of processing pre-digest the meat more, exposing the glutamate and nucleic acids and thus making them taste umame to our tongue.

To compensate for the absence of flavor in our our staches and some of our meats, we now tend to slather them with other flavors, including sweet, salty, acidity and umame from various sauces.

While we appreciate these pleasurable flavors on their own, we like our foods more when these flavors are combined . We like oranges for their combination of sweet and acidic, even as they are mostly devoid of salty and umami; and some oranges are more acidic while others are sweeter. But when all flavors are balanced in one meal, we like our food even more because this indicates that all nutrients are available in somewhat balanced ratios. So we love foods like pizza, sandwiches, hamburgers, stews, and burritos that mix all these flavors together (even though the modern versions of these foods are usually flawed for all the usual reasons). Obviously most chefs know, either technically or intuitively, to combine all these flavors together in their dishes while providing them with certain mouthfeels while also minimizing bitter and astringency. Or in other words, when we consume natural foods: balanced taste equates to optimum taste, which in turn equates to optimum nutrition.

So I recommend that we design our meals to balance the five pleasurable tastes and mouthfeels—that is, sweet, salty, acidic, umami and fatty—while minimizing bitter and astringency—while compensating as much as necessary for the lack of flavor found in our starches and meats. Generally this means starting with the base of your meal—that is, your starch and your protein and from there, adding sweetness, salty and acidity. If your meat lacks umami, then add that with sauces such as miso, soy, cheese and cured meats. Then, as much as necessary, predigest or cook all these foods to break down their bitter and astringent qualities and add the mouthfeel that you want through fat.

PART 4

CHAPTER 31

THE STORY

We can think about diet, perhaps most cogently, in the order that we experience the eating of our food—that is, as one narrative across a timeline. So for that reason, let's consider the case of one single modern man who just read this book: he is still in his twenties, trying to establish his foundational behaviors to make the best of his life. While at the grocery store, he decides he wants to make stew— enough to last him for several meals during his upcoming and busy week. Instead of following any recipe, he wants to improvise, follow his creativity, based on his knowledge of food and nutrients learned in this book.

For animal foods he buys some chicken thighs still covered in skin for their collagen and fat. He could make his own broth but he is pushed for time, so he buys some bone broth instead. Afterwards he selects from the bulk section: pearled barley, lentils, pumpkin seeds, as well as potatoes, yams and an assortment of vegetables with different colors: kale, peppers, carrots, celery, tomatoes and onions.

At home he soaks his pearled barley, lentils and pumpkin seeds in water and apple cider vinegar. The following night after work, he starts to make his stew. In the past he estimated his calories, using the methodology in this book, but now he knows intuitively the amounts of food he needs for the four meals he intends to make from this one stew. He is also able to guess his macronutrients from looking at his proteins and carbs; and he knows that fats are about 14 g per tablespoon. In using his intuition, instead of following any metrics or cookbooks, he feels creative and primal, tapped into something ancient inside of himself.

First, he simmers his mixed vegetables at low temperatures in olive oil and butter, adding salt and pepper and later some herbs from provinces, allowing the aromas to fill his house and quicken his appetite. Once the vegetables are sweated, he notices they do not taste like the vegetables of his childhood that his mother made him eat—not bitter, astringent and weird—but rather sweet, salty, acidic and umami and smooth on his tongue with accentuations of herbs.

He then adds the bone broth, along with some apple cider vinegar and wine for acidity, as well as the diced chicken thighs and skin. From his soak, he adds his pearled barley, red lentils and pumpkin seeds and lets them simmer for about thirty minutes. After tasting the stew, he notes too little umami and adds some miso paste, some more herbs and then lets that simmer down for another thirty minutes.

In this pot, he now has all the nutrients he needs in the ratios in which he needs them—that is, plenty of complete proteins from the chicken and barley, lentils and pumpkin seeds combined; and plenty of collagen protein from the chicken skin and broth and both of those in proper ratios to each other. He also has plenty of carbs from the barley, potatoes, yams, and lentils; plenty of fibers and phytonutrients from various whole plants in his dish; plenty of fat from olive oil, butter and the chicken skin and some additional pop of polyunsaturates, specifically linolenic acid, from the pumpkin seeds. Additionally, his stew smells divine, looks beautiful, full of specks of color from the whole spectrum—greens, yellows, oranges and reds, and glistening with fat while bathing in his yellowish broth.

He calls it the primordial soup.

He pours one serving of his stew into his favorite bowl and decides to eat free of distractions from his phone, etc. He tastes all the flavors on his tongue: the sweet from the yams and vegetables; the acidity from the vegetables, vinegar, and wine; the salt from the minerals; and the umami from the meat, miso, and herbs. He feels the mouthfeel of the fat, smooth on his tongue, not astringent. He notes that nearly all hints of bitter were cooked out of the vegetables which are now soft and sumptuous, while some still have some crunch. He chews the food easily and quickly, noticing how the food lands in his stomach, softly, like it belongs there, as though melting in his gut on contact, as his appetite surges and he spoons down proudly his own, delectable creation while also noticing how soon he seems to reach satiation — as though he hit all the right spots.

And his food does digest more completely and easily for many reasons. It's cut into smaller pieces and thus exposes more surface area to digestive chemicals. His seeds contain fewer antinutrients and toxins from the soaking. And slow cooking at low temperatures predigests food, optimally, while not harming any of the nutrients. The fibers are softened, the proteins denatured, the starches broken down into smaller molecules and gelatinized. Meanwhile, the bitter and astringency are cooked away. While his food digests completely and efficiently, he does not, however, absorb his food too fast into his bloodstream as he would with over-processed foods.

He starts to digest protein in his stomach and then continues the process in his small intestines, releasing amino acids into his bloodstream. Once flowing into his liver, the amino acids then disperse throughout his body, with the complete proteins providing the proper ratio of amino acids to build and maintain his muscles and most of his organs; while the collagen proteins provide for his bones, ligaments, tendons, skin and various parts of his nervous and cardiovascular systems. Because the amino acids are in their proper amounts and ratios, he optimizes his anabolism, avoiding conversions, inefficiencies and wastes.

He starts to digest his carbs in his mouth, which then continues in his small intestine until only the sugars remain, mostly glucose from the starch and some fructose from the yams and vegetables. However, he does not uptake the sugars too quickly in his blood — thus avoiding spikes in his blood sugar and the ensuing detriments. Instead, the carbs are absorbed gradually because of the presence of fiber, protein, fat as well as organic acids; additionally, he chose carbs with different glycemia that raise his blood sugar at different rates: the potatoes absorb the fastest, followed by the barley, legumes and vegetables; finally the fructose from the yams and vegetables is converted into glucose and released into the blood. This, in turn, balances his blood sugar over several hours, allowing for greater mental performance and stability as well as satiation.

At the same time, he only eats enough carbs for the tissues that prefer carbs, mostly his brain and red blood cells. So he does not force feed other tissues glucose or convert any of the glucose into fat—or experience any of the other detriments associated with over-consumption.

He starts to digest his fats in his small intestines until they are just fatty acids and glycerol—which are absorbed into his blood and packaged into lipoproteins. After flowing through the liver, the lipoproteins, along with their fatty acids, are delivered to his cells. Because he consumed his fats in the right amounts, his cells uptake nearly all of them—so they are not processed all over again in his liver or stored as fat. Since the fats are balanced, containing all fatty acids in proper amounts and ratios, the cells easily select the fatty acids needed for their membranes—and then direct the remainder towards catabolism.

The plant fibers are not digested but help him regulate the flow of the food through his gut and finally arrive in his colon. Once there, the soluble fibers ferment, releasing another round of nutrients into his blood: vitamins, minerals, hydrogen gas and the three short-chain fatty acids—all of which provide multiple other benefits.

Additionally, our theoretical consumer absorbs nearly all of the electrolytes in his meal which is abundant in potassium, ample in phosphorous and lower in calcium, magnesium and salt. He does not consume too much salt relative to potassium, or calcium relative to magnesium.

Through eating so many different plants, he also gets an enormous diversity of vitamins and phytonutrients. He acquires plenty of beta carotene for vitamin A from yams and vegetables; and plenty of vitamin K from kale, plenty of vitamin E from his seeds. He gets the full spectrum of plant pigments, including the ones that help his vision and cognition, plenty of organic acids from vegetables, vinegar and wine. He notes that, ever since eating in this way, he rarely gets the sinus and stomach viruses he once experienced.

After his meal is finished, he feels sated and full but not bloated, and feels the warm pulse of energy and electricity in his body. He allows his stew to cool in the pot and then places the whole pot into his refrigerator, saving himself the trouble with extra dishes. Over the next week he spoons out portions of the stew, noting that the stew somehow tastes even better on the second and third time around. In the past, before eating this way, he felt more inclined after dinner to distract himself with video games or some TV series, to escape from himself. But now he feels relaxed and content and decides instead to spend the evening sitting on his porch, listening to music while taking some notes about all the adventures he wants to take in his life. Later he sleeps all the way through the night without waking and feels replenished in the morning.

Because he started this diet somewhat young, he will probably live all or most of his life free of the diseases of civilization while avoiding most infectious diseases and minor ailments and hopefully uses that base to further improve the quality of his life.

CHAPTER 32
PREMISES SUMMARY

PREMISE 1 – DO NO HARM

General rule: Exercise caution about consuming foods and supplements that were not consumed by our ancestors prior to two hundred years ago:

- Avoid artificial and fake ingredients like preservatives, flavorings, colorings, etc.

- Choose organic over conventional foods.

- Choose foods from animals raised as close to their natural habitat as possible.

- Avoid vitamin supplements, unless for some specific and reasoned use. However, vegans should supplement with B12 and possibly others.

- Avoid mineral supplements. Consume salt in moderation, as well as iodine from seaweed and, possibly, clays like bentonite.

- Avoid other supplements unless for some specific or studied purpose.

- Avoid protein powders except, possibly, some forms of collagen.

- Avoid over-processed carbohydrates, like white sugars and breads, etc.

- If you choose not to follow these principles, then do so for particular and intelligent reasons based on research and experimentation, not assumptions or wishful thinking.

PREMISE 2 – REDUCE ANTINUTRIENTS AND TOXINS

General rule: Plants evolved these compounds to protect themselves against predation from other organisms, including humans. Antinutrients block the absorption of nutrients and toxins and harm our tissues. Our ancestors evolved to deactivate these compounds through selecting specific plants and properly processing them through various methods:

- Soak and sprout or ferment your grains—then cook them thoroughly. These methods have various effects upon certain grains, so if needed, conduct further research online.

- Soak and cook your beans, especially if consuming high amounts of them.

- Soak and sprout your nuts, especially if consuming high amounts of them.

- Eat vegetables in moderation, as well as many different types.

- Eat fruit raw or cooked and as much as you want: they do not contain many of these compounds.

- Exercise some caution about your consumption of plant oils.

PREMISE 3 – OPTIMIZE DIGESTION

General rule: When our foods are not completely digested, they flow into our colon and generate detriments; when our foods are not efficiently digested, they increase digestive metabolism and slow us down.

We humans evolved to predigest foods outside of our bodies to reduce our digestive metabolism. Moderns, however, over-process our foods, which strips them of their nutrients and causes us to absorb them too fast. Instead, we should eat whole plants and process them in the proper ways to optimize their nutrition, flavor and satiation:

- Choose to consume more refined but natural and whole foods, such as whole grains, legumes, nuts, tubers, fruits, meats, broths, dairy and eggs.

- When applicable, grind these foods into smaller particles.

- Soak, sprout or ferment your seeds to render their larger macronutrients into smaller macronutrients or even individual nutrients.

- Cook grains and legumes thoroughly.

- Possibly sprout your nuts.

- Eat fruits however you like as they digest easily.

- Eat vegetables in moderation and consider pureeing them

into soups, sauces and the like.

- Consume your meats in different ways: ground, fermented, cured, dried, smoked, soaked in acids, or cooked.

- Ferment some or all of your dairy.

- Consume rendered fats and oils, either raw or cooked.

- Cook most of your food at lower temperatures below 212 degrees: such as steaming, boiling, simmering, confit, etc, to avoid damaging nutrients.

- Consume most of your collagen as broth or finely ground into hamburgers or sausages.

PREMISE 4 – CONSUME ALL NUTRIENTS IN YOUR DIET

General rule: We should consume all nutrients to optimize our metabolism and avoid conversions. Even if just missing one nutrient, we possibly hamper our metabolism.

However, and paradoxically, we also store some nutrients so we can persist for some time without consuming some of them.

We have pools of amino acids that may last for one day or so.

We store enough carbs to last for about one or two days; enough fat to last weeks or months; and enough of some vitamins to last for extended periods of time:

- Consume fatty animal products, including seafood, to ensure you receive complete proteins and all fats in good ratios to each other.

- Consume collagen proteins from broth, etc.

- Consume enough carbs from whole plants, so you also receive their fiber.

- Consume enough additional fat as needed.

- Consume enough fruits and vegetables to attain their special nutrients, including vitamins A, C, K and E, as well as various phytonutrients.

- People who cannot convert enough beta carotene into Vitamin A should possibly consume liver as needed.

- If consuming foods in this way, in these clusters, you indeed receive all nutrients needed—and even if just following your instincts, you probably consume them in about the ratios you need.

PREMISE 5 – OPTIMIZE CALORIES

General rule: We should consume the proper amount of calories to optimize our metabolism. When over-consuming calories, we tend to gain weight regardless of the source of those calories, to some extent anyway. When under-consuming calories, we tend to slow down our metabolism, resulting in fatigue and malaise.

Calories are broken down into four categories of body metabolism: resting, digestive, exercise—and TEE (total energy expenditure), which is the sum of the prior three; TEE is also the total amount of calories you need per day:

- To determine your TEE, you first need to determine your body fat.

- For the best results, use either skin fold calipers, bio-electrical impedance or several others. For next best, use visual body fat estimation, whereby you compare your body to others online.

- Input your data, including your body fat, into the Katch-McArdle Formula—which then calculates your basal while estimating your digestive metabolism; you add your level of exercise and it then calculates the calories you need each day.

- If you do not know your body fat percentage, input your basic data into the Mifflin-St. Jeor Equation or the Institute of Medicine Equation for your results.

- You then need to calculate the number of calories you consume per day by considering nutritional panels and online sources.

- If you are under-consuming calories and feel fatigued or lethargic, try to evaluate the reasons for this:

    o Does your diet contain too much fiber?

    o Too many toxins and antinutrients?

    o Imbalance of nutrients: too much protein relative to fat and carbs? Or too much fat relative to carbs?

- If you are over-consuming and trending towards gaining weight, try to evaluate the reasons:

    o Too many carbs and especially over-processed carbs such as white flours, sugars, juices, etc.

    o Too much fat?

- If you want to lose weight, however, you possibly should reduce your calories. If you want to slow aging (at the expense of slowing your metabolism), you should also reduce calories. (See more on this in the section below.)

- This data is not entirely accurate, probably ranges in accuracy at around 10 to 30%, so use this data only as guidelines, not truths.

- Remember, too, that the more you balance your nutrients, as presented in the next section, the fewer calories you need to consume.

- Ultimately, you should strive not to count calories, but surround yourself with the right food, process that food, and eat that food to balance all your nutrients. Then follow your instincts.

PREMISE 6 —OPTIMUM AMOUNTS AND RATIOS OF NUTRIENTS: PROTEIN (for more information, see the next chapter)

General rule: When we consume macronutrients in their proper amounts and ratios, we then optimize our metabolism—both our anabolism and catabolism, while avoiding inefficiencies and wastes. We should first determine the amount and percentage of protein needed by our bodies, because we do not store much protein in our bodies. When over-consuming protein, we merely convert the amino acids into glucose and ketones. When under-consuming, we pull amino acids from our muscles, thus breaking them down.

Complete proteins are best for anabolizing our muscles and most of our organs. Collagen proteins are best for anabolizing our bones, fascia, ligaments, tendons, skin and some parts of our cardiovascular and nervous systems:

- Consume between 1.0 to 1.6 g of protein per kilogram of weight.

- If sedentary and overweight, choose on the lower end of this range.

- If active and lean, choose on the higher end of this range.

- Consume about 70 to 80% of these proteins as completes; these are complete proteins ranked in their order of quality:

- o   Eggs.

- o   Dairy.

- o   Muscle Meat.

- o   Soybeans.

- o   Combinations of grains and legumes:

- Consume about 20 to 30% of your protein from collagen from such sources as:

  - o   Broth.

  - o   Skin.

  - o   Some ground meats that are mixed with connective tissues.

  - o   Gelatin or collagen supplements (one of my few exceptions to powders or supplements).

  - o   Other foods rich in collagen such as chicken wings (especially when eating the skin, tendons, etc.) and calamari.

PREMISE 6 – OPTIMUM AMOUNT AND RATIO OF CARBS (for more information, see the next chapter)

General rule: We consume carbs to provide glucose for our brain, red blood cells and other tissues — and then we need more for intense forms of exercise. We do not increase the performance of our brain or body by consuming more carbs than this.

When we over-consume carbs, especially refined ones without their fiber, we do not receive any benefits — but only major to minor detriments. When under-consuming carbs, we merely make glucose and ketones in our bodies.

We do not catabolize the sugars, fructose and galactose, which are also more toxic, so we convert them into glucose as quickly as possible. However, fructose provides some additional benefits to our metabolism and blood sugar, especially when attached to fruit or berries.

We store only enough carbs to last about one or two days so we need to consume about that much every day to avoid conversions. We also provide glucose for our brain from other sources—from the conversion of other nutrients. Our brain also catabolizes acetic acid and ketones:

- Consume between 30 and 50% of your calories from carbs. Or somewhere around 150 to 280 carbs per day to provide fuel for the relevant tissues.

- Consume more if you exercise at higher levels of exertion, estimating about 50 carbs per hour.

- Consume most of that as glucose from starch.

- Consume about 10 to 30% of that from fructose.

- Consume nearly all of your carbs from whole plants with their fibers intact, including whole grains, whole tubers, and whole fruits and berries—and grind, pro- cess and predigest them as much as you want.

- If you consume whole carbs, especially in combi- nation with fats, proteins and organic acids, you do not need to worry about high glycemia. However, you may receive benefits from mixing whole carbs that raise your blood sugar at different rates.

Fiber

When consuming whole plants, we automatically consume their fibers as well. These fibers then help regulate the flow of our food through our gut and the uptake of nutrients. Once arriving in our colon, these soluble fibers are fermented, which then releases nutrients into our blood, including vitamins, minerals, phytonutrients, antioxidants and perhaps most importantly the three short-chain fatty acids:

- Eat your carbs whole, and if eating them in the amounts listed above, you should on average consume about 20 to 60 g of fiber per day.

- Eat mixed whole plants, grains, legumes, tubers, fruits and berries, to attain many different types of fibers — which in turn generate greater diversities of nutrients.

- To consume more fermentable fiber, consume more barley, oats, legumes and fruits — or conduct further research.

PREMISE 6 – OPTIMUM AMOUNT AND RATIO OF FATS (for more information, see the next chapter)

General rule: We use fats to make ATP as well as phospholipids for our cell membranes; and we use polyunsaturates, in particular, to make eicosanoids. Under usual conditions, while sleeping, resting, walking or gardening, our muscles and organs prefer to catabolize fats.

When under-consuming fat, we pull from our stores and likely over-consume carbs — which then, paradoxically, possibly results in weight gain. When over-consuming fat, we might suffer detriments; at the same time, we usually do not over-consume fat because we lose our appetite; also the more fat we consume, the more fat we catabolize. (But this is not true for carbohydrates.). In reality, there is little to no correlation between consuming animal and saturated fat and heart disease, assuming you eat these in proper amounts:

- Consume fats at around 30 to 50% of calories.

- Consume enough polyunsaturates from both the omega-6 and -3 families, at around 10 to 30% of total fat.

- Consume more monounsaturates than saturates.

- Consume around two to eight parts omega-6 to one part omega-3.

- Avoid combining high amounts of fats and carbs in the same meal, as the carbs block the utilization of fats, causing detriments.

- Mix your types of fats between animals and plants.

  o Use animal fat as the base of your diet, as found in tallow, eggs, dairy and meat.

  o Add plant oils low in omega-6, linolenic acid, including olive, avocado, peanut and several other nut oils.

  o Eat other plant oils in moderation or not at all.

  o Eat nuts and fatty seeds regularly.

  o Eat some foods high in omega-3, like seafood or small amounts of specific seeds like flax, chia and hemp, as well as dairy, eggs and flesh from pastured or wild animals.

PREMISE 7 – TASTE AND SATIATION

General rule: Your diet must taste delicious and be satiating; otherwise, you will not adhere to that diet.

Humans have four tastes that are correlated to nutrients we need: umami for protein; sweet for carbs; acidic for organic acids; salty for sodium, chloride and potassium. We have mouthfeel for fat. And bitter for toxins—which warns us to avoid some foods.

When combining all these flavors together with natural foods, we thus combine our nutrients to optimize our metabolism. So to some extent, the tastier our food, the more nutritious, as long as we adhere to the principles of this diet:

- Learn to identify these flavors.

- Choose meals that conform to your taste and nutritional goals.

- Learn to cook to optimize these flavors and mouthfeels, as well as your nutrition.

- Consider slow cooking in ceramic pots as one of the best and most ancestral ways.

MACROMINERALS

General rule: Humans utilize six different electrolytes: potassium, magnesium, phosphorous, calcium, sodium, and chloride to control the flow of liquids and electricity in our bodies.

Electrolytes are atoms, not molecules, so they are not broken down into smaller units which means we can recycle and reuse them endlessly, provided they are not eliminated. We regulate them in our blood and other tissues by taking up only certain amounts and then retaining or expelling certain amounts in our kidneys. We probably cannot over-consume electrolytes through food, but we can as supplements, including salt; we possibly can overconsume calcium and phosphorous from too much dairy. We possibly can under-consume electrolytes, but doubtful. However, we can easily imbalance them, leading to considerable detriments:

- Eat this diet as recommended and you automatically receive enough electrolytes.

- Eat salt in small or moderate amounts.

- Avoid mineral supplements, except for iodine.

- Possibly avoid consuming too much calcium through dairy in relationship to magnesium.

- If you do not consume dairy, you probably do not need to supplement with calcium—but if you do, use some form of ground bone or whole bones from small animals like sardines.

PHYTONUTRIENTS

Various phytonutrients are associated with these benefits: anti-microbial, antioxidants, anti-inflammatory, antiallergic, anti-spasmodic, anti-cancer, anti-aging, hepatoprotective, hypolipidemic, neuroprotective, hypotensive, diabetes, osteoporosis, CNS stimulant, analgesic, protection from UVB-induced carcinogenesis, immuno-modulator, and carminative.

Certain plants contain certain phytonutrients that provide various benefits to our bodies:

- Eat the mixture of plant foods contained in this diet.

- Eat at least ten different plants per day.

- Eat plants across the spectrum of colors.

- Eat plenty of sours and tarts in your diet.

- Eat fermented foods.

- Consider eating certain plants raw, especially fruits and leaves, as they contain greater amounts of phytonutrients undamaged by cooking.

VEGETARIANISM AND VEGANISM

General rule: On this diet, you can easily eat vegetarian while still attaining the proper amounts and ratios of all, or nearly all, nutrients:

- Replace meat with eggs and dairy, which are better sources of protein anyway; they also provide additional nutrients not found in meat, such as vitamin A.

- Replace meat also with combinations of grains, legumes and nuts.

- You cannot acquire collagen from eggs and dairy or plants. However, eggs and dairy probably generate collagen better in our bodies than most forms of meat. Otherwise eat foods high in vitamin C to boost collagen production insider your body.

- If you cannot convert short-chain omega-3s into their longer chains, you may want to focus on some source of EPA and DHA, perhaps from algae or pastured eggs.

You can somewhat easily eat Vegan in this diet through carefully combining plants to make complete proteins, namely grains, legumes and nuts—and then properly processing those foods:

- You can acquire your saturated fats from red palm oil or cacao butter.

- As noted, you need to supplement with vitamin B12, possibly other vitamins, as well as vitamin A (if you cannot convert enough beta carotene)

CHAPTER 33

SUMMARY OF MACRONUTRIENTS

RANGE OF NUTRIENTS FOR OUR THEORETICAL CONSUMER

Theoretical consumer: Thirty-five years old, either male or female, somewhat fit and active.
2300 daily calories based on calorie calculators.

PROTEIN: 10 to 15% of calories (58 to 82 g)
Completes: 70 to 80% of total protein from dairy, eggs, meats and combinations of grains and legumes and/or nuts.
Collagen: 20 to 30% of total protein from broth, skins, chicken, sausage and possibly some powders.

CARBS: 35 to 50% of total calories (200 to 288 g)
Glucose: 70 to 90%.
Fructose: 10 to 30%.

- Whole carbs from grains, tubers, legumes, fruits and berries; some honey is ok.

- Properly process most carbs, especially grains and legumes.

- When eating carbs in this way, you naturally consume between 20 to 50 g of diverse fibers and plentiful phytonutrients.

FATS: 35 to 50% of total calories (89 to 128 g)
Saturated: 30 to 40% of total fat.
Monounsaturated: 40 to 50% of total fat.
Polyunsaturated: 10 to 30% of total fat. Around four to ten parts omega-6 to one part omega-3.

- Properly process most nuts.

- Consume fats cooked or raw.

OPTIMUM RATIO OF NUTRIENTS FOR OUR THEORETICAL CONSUMER

Theoretical consumer: 35 years old, fit and active.

PROTEIN: 15% of total calories (82 g)
Completes: 70 to 80% of total protein.
Collagen: 20 to 30% of total protein.

CARBS: 45% of total calories (258 g)
Glucose: 80%.
Fructose: 20%.

- Whole carbs from grains, tubers, legumes, fruits and berries; some honey is ok.

- Properly process most carbs, especially grains and legumes.

- 25 to 50 g of fiber.

FATS

Fats: 40% of total calories (102 g)

Saturated: 30% of total fat.

Monounsaturated: 50% of total fat.

Polyunsaturated: 20% of total fat. Around six parts omega-6 to one part omega-3.

Properly process most nuts.

Consume fats cooked or raw.

Theoretical Consumer: 2300 Calories per day

Type	Catagory	Optimum Percentages	Optimum G	Range Percentages	Range G
Protein	Total protein	15%	82	10%-15%	58-82
	Complete	70%-80%	57-66	70%-80%	57-66
	Collagen	20%-30%	16-25	20%-30%	16-25
Carbs	Total Carbs	45%	258	35%-50%	200-288
	Total Glucose	80%	206	70%-90%	202-259
	Total Fructose/ Galactose	20%	52	10%-30%	29-86
	Total Fiber		35		25-50
	Soluble		18		13-25
	Insoluble		18		13-25
Fats	Total Fats	40%	102	35%-50%	89-128
	Saturated	30%	31	30%-40%	38-51
	Monounsaturated	50%	51	40%-50%	51-64

Polyunsat-urated	20%	20	10%-30%	13-38
Omega 6 to 3 ratio	6:1		2 to 10:1	

TABLE 33.1- RANGE AND OPTIMUM PERCENTAGES OF MACRONUTRIENTS FOR OUR THEORETICAL CONSUMER WHO IS THIRTY-FIVE YEARS OLD, NEITHER MALE OR FEMALE, SOMEWHAT FIT AND ACTIVE

NOTES

- The better you balance the nutrients in your diet, the fewer nutrients you need. So you possibly need way fewer calories than suggested here—but about the same ratio of nutrients.

- Nutrients, obviously, work in balance with each other. In some cases, if you are deficient or excessive in any one nutrient, you might imbalance this whole system.

- Supplement with iodine from seaweed but at the recommended amounts.

- Some people cannot convert enough of the shorter-chain omega-3s, ALA, into the longer chains, EPA and DHA. In that case, you should eat some seafood every week.

- Some people also cannot convert enough beta carotene into Vitamin A; so you should eat liver, especially if not eating dairy and eggs (which contain some vitamin A) or possibly supplement with vitamin A.

- Generally, cook most of your foods at lower temperatures to optimize pre-digestion while not harming nutrients. Avoid ultra-pasteurized or UHT products from the grocery store, including some dairy and most boxed or canned foods.

- If concerned about uric acid, inflammation and gout, consume more of your proteins from dairy or eggs, which are low in purines.

- Consume some raw fruits and vegetables that are high in vitamin C, which is damaged during cooking and even freezing.

- Remember this diet is to optimize your metabolism — and thus also your biological potential. However, if you want to apply this diet to lose weight, try these recommendations in this order:

  o Eat this diet as recommended and see if you start to lose weight without any additional effort.

  o Perform low to moderate forms of exercise as part of your regular routine: walk to work, hike on the weekends with friends, clean the house, work in the yard. At this rate, you catabolize about 300 calories per hour, which over seven days is one pound of body fat.

  o Avoid entirely any over-processed carbs and even consider under-processing them: eating whole grains, in larger pieces, such as rolled or cut oats and intact beans.

  o Eat foods higher in fiber and lower in macronutrients, including more legumes, vegetables and fruits as well as whole grains higher in fiber, including barley and oats.

  o Eat more of your fat attached to whole plants, such as avocados and nuts.

  o Consume lower amounts of carbs under 200 g. Under this level they are catabolized, not stored.

  o Possibly trend towards eating more protein to induce

satiation sooner, quell appetite and lower insulin.

o   If this does not work naturally, then lower the amount of fat in your diet, keep carbs under 180 g or lower, and keep protein about the same or raise slightly and exercise more at low to moderate levels of exercise.

ANTI-AGING

If you are trying to delay or reverse aging, eat this diet as recommended. Through balancing your nutrients you need fewer of them overall, which in turn reduces metabolism in some parts of your body—thus slowing aging.

- Periodically fast to induce autophagy to remove cellular junk.

If you want to delay aging even more, you probably should consume fewer calories, but at the same time you need to reduce your metabolism. So, reduce protein to slow protein turnover. Lower fat to reduce basal by only as much as 10%. Possibly reduce carbs. And reduce digestive and exercise metabolism.

AFTERWORD

With this book I hope I have helped you understand your diet and given you the tools to apply that understanding to enhance the quality of your life, avoid the diseases of civilization as well as many infectious diseases and other ailments as well. In the application of this diet, you will probably take different approaches. If your personality is more regulated and structured, you may choose to follow all the metrics. If more conceptual and intuitive, you may choose to grok the basic ideas and flow with that. Both, of course, will work.

But in either case, you should strive, in the end, to make your diet more intuitive by following these simple steps: eliminate bad foods, embrace good foods, and then prepare and combine those foods as outlined in this book.

Then, just trust your gut. Though only mentioned in passing in this book, you actually have feedback loops, located in your gut, liver, brain and hormones, that guide you towards the foods and nutrients you need. They not only guide you towards the amount of calories but also towards specific nutrients. So, if you control your culinary environment as mentioned above, you can likely just trust that you will fall in alignment with your optimum diet—even though, in our modern world, you can easily lose your way just with one or two bad meals or snacks.

You should also trust your gut for another reason. While you now understand most aspects of nutrition, scientists still collectively do not understand nutrigenetics all that well—that is, how you are possibly adapted to specific foods and nutrients due to your unique ancestry or other factors. If your ancestors were Inuit or Plains Indians, adapted to diets of fatty animals, you may suffer your whole life because you are eating too many carbs, plant oils, dairy products, and salt. If your ancestors evolved for diets rich in omega three, you may suffer from too many omega sixes and etc. For this reason, you should listen to your gut even more because it may tell you, in one way or another, that you are not adapted for some specific food or for greater or lesser amounts of various nutrients.

You can also work around nutrigenetics by conducting as much research as possible about the diets of your most recent ancestors—that is, consider the foods and nutrients they consumed and then research if any of those foods are affected by nutrigenetics and how. However, not much information is available here. To complicate matters, most of us are now mixed breeds with our ancestors in our lineage eating many different types of diets. As noted before, nearly all people with European genes are descended from three different foodways: herders, farmers and foragers.

You can also work around nutrigenetics by using other various techniques pertaining to specific foods or nutrients as listed below:

- We all need ample amounts of vitamin A, but some of us can convert enough beta carotene into vitamin A and others cannot. In any case, though, if we all eat enough preformed vitamin A from eggs, dairy and liver, then none of us will suffer defecencies of vitamin A.

- Europeans evolved to consume more salt and thus eliminate it better. Africans evolved to consume less and retain it better. However, in all cases, nobody needs to consume over 500 to 1000 mg per day.

- Some people are better adapted to consume less vitamin C, some more, but nobody evolved to experience any detriments from consuming enough vitamin C. So everyone should always consume enough vitamin C.

- Some people are better adapted to alcohol; some less. In this particular case, you should listen to your body. And if you chose to drink, then do so in moderation.

- Some people are better adapted to greater amounts of carbs relative to fats, explaining why I created the range in these macronutrients. On this one, listen to your body.

- Some people are likely better adapted to certain fatty acids. Listen to your body.

- Some people are allergic to some foods or more sensitive to some secondary plant compounds or storage proteins like gluten. Listen to your body.

- You can easily research other forms of nutrigenetics.

As mentioned before, scientists do not yet really understand the effects of epigenetics on diet. Epigenetic refers to molecules that form from cues in our environment that then tell which genes to turn on or off, or to what extent. Oddly enough, our parents can develop these molecules—and hand them down to their offspring. Our epigenetics appears to change in relationship to our diet—but scientists do not understand this much at this point. So again, you should listen to your gut.

As mentioned in the Introduction, diet, more than anything, will determine your health throughout your life. But other basic factors contribute to your health as well. In the first book of the series, On the Origins of Being, I and my co-author, Jenny Powers, discuss these basic factors, including:

- Circadian rhythms, which include sleep, waking, work and rest. These rhythmns are set mostly by our exposure to natural or outdoor light.

- Exposure to nature.

- Level and extent of exercise.

- And range of motion.

In addition to these basic needs, we also have social and cultural needs—which books two and three cover in that series. We prove that when living in alignment with our evolutionary needs, we are way more likely to experience health and well-being; and when in mismatch, pysical and mental pathology. So I hope you continue your journey towards fulfilling your human potential.

EPISTEMOLOGY AND TERMINOLOGY

How do we know what we know?

Many people claim to know things—but how often do we ask: how do they know?

You should ask this question of me, the author.

As already presented, I claim to know about diet via the four methodologies that are generally considered the best way to understand nutrition—along with the fifth methodology that I mentioned in my "Afterword": trusting your own gut? On top of that, I then compare these four methodologies to each other and reconcile the discrepancies—with all of this working, more synergistically, to reveal the truth on what is, in reality, one of the most complicated subjects known—that is, how hundreds of nutrients work with each other to optimize our health.

But within this framework: how do I know what I know, especially given that these four methodologies are super complicated and esoteric? Nutritional physiology covers how invisible atoms and molecules behave in our body. Ancestral consumption covers how our ancestors evolved over millions of years, even though they are now extinct and we cannot observe them, and we only have limited clues. The other two methodologies also have their complications. So how do I know what I know in this case?

Evidence, of course, is the answer.

Within these methodologies, I take different approaches to evidence, some more strict, others more lenient. With nutritional physiology, I rely mostly on putative, textbook science that is mostly unchallenged over many decades now. For this reason, I usually do not footnote this stuff because it is easily accessible via AI, Wikipedia, internet searches and textbooks; you can then cross-reference that data for accuracy, too. When this data is more controversial or recent, I footnote it, but of course, there is considerable room for discretion here. To keep this book from getting too fat, and scaring away my readers, I was prone towards footnoting less rather than more.

With ancestral consumption, I take another approach with the evidence. First of all, the subject is more esoteric and less putative. But instead of footnoting more, I instead rely upon my previous work, the series of books on the subject, called "The First Supper" which is rich with citations. But here I am taking enormous amounts of data from those books, forming them into one larger theory and narrative and then connecting many dots in between and, for many reasons, this proves difficult to footnote.

In most cases I do not footnote establishment recommendations because that data is easy to find. But I almost always footnote scientific studies with reputable sources. In some cases, I may take some liberties by asserting something that is in all probability true, but not necessarily proven or sourced as true. I usually footnote to links on the internet as much as possible, so my reader can easily access the source—even while acknowledging that the best data is frequently behind pay and subscription walls.

I also use another epistemology in this book—that is, narrative. For nutritional physiology, I tell the whole, albeit condensed, narrative of nutrients and their function inside our bodies in chronological order. I do the same for hominoids in ancestral consumption. Through telling these stories, and then analyzing these stories from various scientific desciplines, I better elucidate the underlying patterns and connections in the data to reveal larger truths—and make the data more grokable to my reader. In the process, I connect some dots along the way and sometimes jump over dots.

I also take considerable liberties with my scientific terminology but for deliberate reasons—that is, to streamline my writing to make this book more accessible to more readers. For example I write something like this: "fats are anabolized into phospholipids," which is problematic for several reasons. First, I turn the word "anabolism" into the verb that does not exist, "anabolized." Second, scientists tend not to apply the term anabolism to fats; instead they use the term synthesized. Furthermore, in making that statement, I am removing so many chemical steps that one might argue the statement is not true. But ultimately, in the reality of biology, fatty acids are indeed eventually anabolized (built into larger molecules) into phospholipids inside our bodies—and that is what I want the reader to know. I take similar liberties throughout the book but only to serve my reader with the key information in simple terms.

With all this said, I always keep in mind that evidence is slippery, even when footnoted from the best sources. Though science is the best way to determine much of the truth of our lives, it also is not perfect, and like every other part of our lives, ample with self, financial and ideological interests that can result in flawed evidence. Even when free of all that, scientists, like any human, are subject to making mistakes and bad judgment or, otherwise, just groping into the mystery, half confused, even with brilliant and honest minds. With that being said, I hope my reader realizes that I struggled harder with the truth of this subject more than most or even all, while acknowledging the frailty of knowledge—which, paradoxically, better helps me find the truth. With this book, I am not striving for perfection but usefulness: to write the most advanced and comprehensive, and carefully considered, book on this subject yet to date and to make this book accessible to larger audiences, so they can use the knowledge for their own benefit.

Endnotes

1. Wikipedia. 2025. "Systems science." Accessed September 2, 2025. https://en.wikipedia.org/wiki/Systems_science#:~:text=System%20 dynamics%20is%20an%20approach,loops%20and%20stocks%20 and%20flows.

2. Westerterp, K.R. 2004. "Diet induced thermogenesis." Nutr Metab 1(5).

3. Popova, A., and Mihaylova, D. 2019. "Antinutrients in Plant-based Foods: A Review." The Open Biotechnology Journal 13: 68.

4. Kocyigit, E., Kocaadam-Bozkurt, B., et al. 2023. "Plant Toxic Proteins: Their Biological Activities, Mechanism of Action and Removal Strategies." Toxins (Basel) 15(6):356.

5. Biesiekierski, J.R. 2017. "What is gluten?" J Gastroenterol Hepatol 32(S1):78-81.

6. Wikipedia. 2025. "Lifestyle disease." Accessed September 2, 2025. https://en.wikipedia.org/wiki/Lifestyle_disease.

7. Ajomiwe, N., Boland, M., et al. 2024. "Protein Nutrition: Understanding Structure, Digestibility, and Bioavailability for Optimal Health." Foods 13(11):1771.

8. Bionumbers. "Percent of protein in body that is collagen." Accessed September 2, 2025. https://bionumbers.hms.harvard.edu/bionumber.aspx?id=109730.

9. Schutz, Y. 2011. "Protein turnover, ureagenesis and gluconeogenesis." Int J Vitam Nutr Res 81(2-3):101-7.

10. "6.4: Protein Digestion and Absorption." Accessed September 2, 2025. https://med.libretexts.org/Bookshelves/Nutrition/An_Introduction_to_Nutrition_(Zimmerman)/06%3A_Proteins/6.04%3A_Protein_Digestion_and_Absorption.

11. Schutz, "Protein turnover, ureagenesis and gluconeogenesis."

12. Molnar, C., and Gair, J. 2015. "Nitrogenous Wastes." Concepts of Biology – 1st Canadian Edition. Victoria, British Columbia: BC Campus/Creative Commons.

13. Harvard Health. 2011. "Abundance of fructose not good for the liver, heart." Accessed September 2, 2025. https://www.health.harvard.edu/heart-health/abundance-of-fructose-not-good-for-the-liver-heart.

14. Wikipedia. "Gluconeogenesis." Accessed September 2, 2025. https://en.wikipedia.org/wiki/Gluconeogenesis.

15. Oliphant, K., and Allen-Vercoe, E. 2019. "Macronutrient metabolism by the human gut microbiome: major fermentation by-products and their impact on host health." Microbiome 7 (91).

16. Oliphant, "Macronutrient metabolism by the human gut microbiome."

17. Waniewski, R.A., and Martin, D.L. 1998. "Preferential utilization of acetate by astrocytes is attributable to transport." J Neurosci 18(14):5225-33.

18. Saghafian, F., Hajishafiee, M., et al. 2023. "Dietary fiber intake, depression, and anxiety: a systematic review and meta-analysis of epidemiologic studies." Nutr Neurosci 26(2):108-126.

19. Yang, M., Cai, C., et al. 2024. "Effect of dietary fibre on cognitive function and mental health in children and adolescents: a systematic review and meta-analysis." Food Funct 15(17):8618-8628.

20. Timm, M., Offringa, L.C., et al. 2023. "Beyond Insoluble Dietary Fiber: Bioactive Compounds in Plant Foods." Nutrients 15(19):4138.

21. Roy, D. 2024. "11.5: Neutralization of Fatty Acids and Hydro-

lysis of Triglycerides." Accessed December 9, 2024. https://chem.libretexts.org/Courses/American_River_College/CHEM_309%3A_Applied_Chemistry_for_the_Health_Sciences/11%3A_Lipids_-_An_Introduction/11.05%3A_Neutralization_of_Fatty_Acids_and_Hydrolysis_of_Triglycerides.

22. Zhuang,Y., Dong, J., et al. 2022. "Impact of Heating Temperature and Fatty Acid Type on the Formation of Lipid Oxidation Products During Thermal Processing." Front Nutr 2(9): 913297.

23. Lehner, R., and Quiroga, A.D. 2016. "Fatty Acid Handling in Mammalian Cells." In Biochemistry of Lipids, Lipoproteins and Membranes, edited by Neale D. Ridgway and Roger S. McLeod. Amsterdam, Netherlands: Elsevier.

24. Tsaluchidu, S., and Puri, B.K. 2008. "Fatty acids and oxidative stress." Ann Gen Psychiatry 7(Suppl 1):S86.

25. Gimenez, M.S., Oliveros, L.B., and Gomez, N.N. 2011. "Nutritional deficiencies and phospholipid metabolism." Int J Mol Sci 12(4): 2408-33.

26. Wall, R., Ross, R.P., et al. 2010. "Fatty acids from fish: the anti-inflammatory potential of long-chain omega-3 fatty acids." Nutr Rev 68(5): 280-9.

27. Tannock, G.W., and Liu, Y. (2019). "Guided dietary fibre intake as a means of directing short-chain fatty acid production by the gut microbiota." Journal of the Royal Society of New Zealand 50(3), 434–455.

28. Vockley J. 2020. "Long-chain fatty acid oxidation disorders and current management strategies." Am J Manag Care 26(7 Suppl):S147-S154.

29. Heymsfield, S.B., Gonzalez, M.C., et al. 2014. "Weight loss composition is one-fourth fat-free mass: a critical review and critique of this widely cited rule." Obes Rev 15(4):310-21.

30. Higdon. J. 2025. "Essential Fatty Acids." Accessed September 2, 2025. https://lpi.oregonstate.edu/mic/other-nutrients/essential-fatty-acids.

31. World Health Organization. 2024. "Fat Intake." Accessed December 9, 2024. https://www.who.int/data/gho/indicator-metadata-registry/imr-details/3418#:~:text=Excessive%20dietary%20fat%20intake%20has,the%20shift%20in%20consumption%20patterns.

32. Raichlen, D.A., Pontzer, H., et al. "Sitting, squatting, and the evolutionary biology of human inactivity." Proc. Natl. Acad. Sci. U.S.A. 117(13): 7115-7121.

33. Steegmann. A.T. 2007. "Human cold adaptation: An unfinished agenda." Am J of Human Biol 19(2):218-27.

34. Wikipedia. 2025. "Punctuated equilibrium." Accessed September 2, 2025. https://en.wikipedia.org/wiki/Punctuated_equilibrium.

35. Altmann, S.A. 2009. "Fallback foods, eclectic omnivores, and the packaging problem." Am J Phys Anthropo 140(4): 615-29.

36. Lorenzo, P.M., Izquierdo, A.G., et al. 2022. "Epigenetic Effects of Healthy Foods and Lifestyle Habits from the Southern European Atlantic Diet Pattern: A Narrative Review." Adv Nutr 13(5): 1725-1747.

37. Carneiro, V.C., and Lyko, F. 2020. "Rapid Epigenetic Adaptation in Animals and Its Role in Invasiveness." Integr Comp Biol 60(2): 267–274.

38. Jacobs, I., and Osvath, M. 2016. "Nonhuman Tool Use." In Encyclopedia of Evolutionary Psychological Science, edited by V. Weekes-Shackelford, and T. Shackelford. Cham, Switzerland: Springer.

39. Almécija, S. 2016. "Pitfalls reconstructing the last common ancestor of chimpanzees and humans." Proc Natl Acad Sci U.S.A.113(8): E943-E944.

40. Gerstner, K.F., and Pruetz, J.D. 2022. "Wild Chimpanzee Welfare: A Focus on Nutrition, Foraging and Health to Inform Great Ape Welfare in the Wild and in Captivity." Animals (Basel) 12(23): 3370.

41. Milton, K. 2000. "Hunter-gatherer diets—a different perspective."

Am J Clin Nutr 71(3): 665-667.

42.   Stanford, C.B., and Nkurunungi, J.B. 2003 "Behavioral Ecology of Sympatric Chimpanzees and Gorillas in Bwindi Impenetrable National Park, Uganda: Diet." Int J Primatol 24(4): 901–918.

43.   Kamel, K.S., and Halperin, M.L. 2017. "Chapter 6 – Metabolic Acidosis: Acid Gain Types." In Fluid, Electrolyte and Acid-Base Physiology (Fifth Edition)., edited by Kamel S. Kamel, Mitchell L. Halperin. Amsterdam, Netherlands: Elsevier. p 141-170.

44.   Wade, L. 2016. "How sliced meat drove human evolution." Accessed December 11, 2024. https://www.science.org/content/article/how-sliced-meat-drove-human-evolution#:~:text=The%20most%20tedious%20part%20of,jaws%20our%20early%20ancestors%20had.

45.   University of Cambridge. 2025. "Genetic study reveals hidden chapter in human evolution." Accessed September 2, 2025. https://phys.org/news/2025-03-genetic-reveals-hidden-chapter-human.html.

46.   Yilmaz, F., Karageorgiouet, C., et al. 2024. "Reconstruction of the human amylase locus reveals ancient duplications seeding modern-day variation." Science 386(eadn0609).

47.   Sepulchre, P.R., Flatueu, G., et al. 2006. "Tectonic Uplift and Eastern Africa Aridificaiton." Science 313: 1419-1423.

48.   Sheriff, M. 2017. "Born to cope with stress." Accessed December 11, 2024. https://researchfeatures.com/born-to-cope-with-stress/.

49.   Hernandez-Aguilar, R.A., Moore, J., et al. "Savanna chimpanzees use tools to harvest the underground storage organs of plants." Proc Natl Acad Sci U.S.A.104 (49):19210-19213.

50.   Callaway, E. 2016. "Meet Chewie, the Biggest Australopithecus on Record." Accessed December 11, 2024. https://www.scientificamerican.com/article/meet-chewie-the-biggest-australopithecus-on-record1/.

51.   Nishida, T. 2024. "Chimpanzee." Accessed December 11, 2024.

https://www.ifaw.org/animals/chimpanzees.

52. Tuttle, R.H. 2025. "Theories of bipedalism." Accessed September 2, 2025. https://www.britannica.com/science/human-evolution/Theories-of-bipedalism.

53. Milton, K. 2003. "The Critical Role Played by Animal Source Foods in Human (Homo) Evolution." J Nutr 133(11): 3886S-3892S.

54. Uy, J., Beresnevi i t , G., and Nguyen, V. 2024. "The Relationship of the Lower Ribcage with Liver and Gut Size: Implications for Paleoanthropology." Humans 4(4): 310-320.

55. Ward, C.V., and Hammond, A.S. 2016. "Australopithecus and Kin." Nature Education Knowledge 7(3):1.

56. Pobiner, B. 2013. "Evidence for Meat-Eating by Early Humans." Nature Education Knowledge 4(6):1.

57. Wells, W.A. 2007. "Big mammals have big (or slow) cells." J Cell Biol 176(7): 893.

58. Talal, S., Harrison, J.F., et al. 2024. "Body mass and growth rates predict protein intake across animals." eLife 13: e88933.

59. Hendry, L. ND. "Human Evolution: Homo erectus, our ancient ancestor." Accessed December 11, 2024. https://www.nhm.ac.uk/discover/homo-erectus-our-ancient-ancestor.html#:~:text=Adults%20grew%20to%20about%201.4,and%20their%20skulls%20were%20thicker.

60. Carroll, S.B. 2003. "Genetics and the making of Homo sapiens." Nature 422 (6934): 849-857.

61. Bramble, D.M., and Lieberman, D.E. 2004. "Endurance running and the evolution of Homo." Nature 432 (7015): 345-352.

62. Wrangham, Richard W. 2009. Catching Fire: How Cooking Made Us Human. New York: Basic Books.

63. Steudel-Numbers, K.L. 2006. "Energetics in Homo erectus and other early hominins: the consequences of increased lower-limb length." J Hum Evol 51 (5): 445-453.

64. Wang, Z., Ying, Z., et al. 2011. "Evaluation of specific metabolic

rates of major organs and tissues: comparison between men and women." Am J Hum Biol 23(3): 333-8.

65. Carroll, "Genetics and the Making of Homo sapiens."

66. Smithsonian National Museum of Natural History. 2024. "What does it mean to be human? Human Skin Color Variation." Accessed December 11, 2024. https://humanorigins.si.edu/evidence/genetics/human-skin-color-variation#:~:text=Melanin%2C%20the%20skin's%20brown%20pigment,changes%20that%20improve%20reproductive%20success.

67. Smithsonian National Museum of Natural History. 2022. "What does it mean to be human? Bodies: Changing Body Shapes and Sizes." Accessed December 11, 2024. https://humanorigins.si.edu/human-characteristics/bodies.

68. Creanza, N., Kolodny, O., et al. "Cultural evolutionary theory: How culture evolves and why it matters." Proc Natl Acad Sci U.S.A. 114 (30):7782-7789.

69. Alex, B. "How Did Ancient People Keep Their Food From Rotting?" Accessed December 12, 2024. https://www.discovermagazine.com/planet-earth/how-did-ancient-people-keep-their-food-from-rotting.

70. Isler, K. 2016. "Metabolic Acceleration in Human Evolution." Cell Metab 24(1): 5-6.

71. Perkins, J.M., Subramanian, S.V., et al. 2016. "Adult height, nutrition, and population health." Nutr Rev 74(3): 149-65.

72. Schwarz, J. 2003. "Bones from French Cave Show Neanderthals, Cro-Magnon Hunted Same Prey." Accessed February 2, 2025. https://www.washington.edu/news/2003/09/22/bones-from-french-cave-show-neanderthals-cro-magnon-hunted-same-prey/.

73. Kelly. R.L. 2013. The Lifeways of Hunter Gatherers: The Foraging Spectrum. Cambridge, England: Cambridge University Press.

74. Cordain, L., Miller, J.B., et al. 2000. "Plant-animal subsistence ratios and macronutrient energy estimations in worldwide hunter-gatherer diets." Am J Clin Nutr 71 (3): 682-692.

75. Krogh, A., and Jorgensen-Krough, M. 1915. A Study of the Diet

and Metabolism of Eskimos Undertaken in 1908 on an Expedition to Greenland. Denmark: Kommission Hos C.A. Reitzel.

76.   Comer, L. "Inuit Foodways." Accessed September 3, 2025. https://www.thefirstsupperbooks.com/wp-content/uploads/2019/06/Inuit-Foodways.pdf.

77.   White, T., Asfaw, B., et al. 2003. "Pleistocene Homo sapiens from Middle Awash, Ethiopia." Nature 423: 742–747.

78.   McDougall, I., Brown, F., et al. 2005. "Stratigraphic placement and age of modern humans from Kibish, Ethiopia." Nature 433: 733–736.

79.   Wikipedia. 2025. "Klasies River Caves." Accessed September 3, 2025. https://en.wikipedia.org/wiki/Klasies_River_Caves.

80.   Wikipedia. 2025. "Kebaran culture." Accessed September 3, 2025. https://en.wikipedia.org/wiki/Kebaran_culture.

81.   Wikipedia. 2025. "Natufian culture." Accessed September 3, 2025. https://en.wikipedia.org/wiki/Natufian_culture.

82.   Naithani, S. 2021. "The Origins of Agriculture." In History and Science of Cultivated Plants. Creative Commons: Oregon State University.

83.   Cooper, T. 2024 "A Brief History of Goat Domestication." Accessed December 12, 2024. https://livestockconservancy.org/2022/10/11/a-brief-history-of-goat-domestication/.

84.   Katz, D.C., Grote, M.N., et al. 2017. "Changes in human skull morphology across the agricultural transition are consistent with softer diets in preindustrial farming groups." Proc Natl Acad Sci U.S.A. 114(34): 9050-9055.

85.   Egan, D. "Neoteny, Play, and the Anthropological Difference." Accessed September 3, 2025. https://sites.cisa-unige.ch/ronald-de-sousa/assets/pdf/Egan_Paper.pdf.

86.   Mercader, J. 2009. "Mozambican Grass Seed Consumption During the Middle Stone Age." Science 326: 1680-1683.

87.   Luca, F., Perry, G.H., et al. 2010. "Evolutionary adaptations to dietary changes." Annu Rev Nutr 30:291-314.

88. Dudley, R., and Maro, A. 2021. "Human Evolution and Dietary Ethanol." Nutrients 13(7):2419.

89. Hockings, K.J., Bryson-Morrison, N., et al. 2015. "Tools to tipple: ethanol ingestion by wild chimpanzees using leaf-sponges." R Soc Open Sci 2(6):150150.

90. Itan, Y., Powell A., et al. 2009. "The Origins of Lactase Persistence in Europe." PLoS Comput Biol 5(8):e1000491.

91. Sissons, B. 2023. "What to know about milk and gout." Accessed December 13, 2024. https://www.medicalnewstoday.com/articles/milk-and-gout.

92. Ryan, K., Fitts, W., et al. 2009. "Tracking East African Cattle Herders from Prehistory to the Present." Expedition Magazine 51(3): 27-36.

93. Page, A.E., Viguier, S., et al. 2016. "Reproductive trade-offs in extant hunter-gatherers suggest adaptive mechanism for the Neolithic expansion." Proc Natl Acad Sci U S A 113 (17): 4694-4699.

94. Lazaridis, I., Patterson, N., et al. 2014. "Ancient human genomes suggest three ancestral populations for present-day Europeans." Nature 513: 409–413.

95. Corbett, S., Courtiol, A., et al. 2018. "The transition to modernity and chronic disease: mismatch and natural selection." Nat Rev Genet 19: 419–430.

96. Comer, L., and Powers, J. 2024. On the Origin of Being. Austin, TX: River Grove Books.

97. Mariamenatu, A.H., and Abdu, E.M. 2021. "Overconsumption of Omega-6 Polyunsaturated Fatty Acids (PUFAs) versus Deficiency of Omega-3 PUFAs in Modern-Day Diets: The Disturbing Factor for Their 'Balanced Antagonistic Metabolic Functions' in the Human Body." J Lipids 2021:8848161.

98. Calcagno, M., Kahleova, H., et al. 2019. "The Thermic Effect of Food: A Review." J Am Coll Nutr 38: 1-5.

99. Carmody, R.N., Weintraub, G.S., et al. 2011. "Energetic consequences of thermal and nonthermal food processing." Proc Natl

Acad Sci U S A 108(48):19199-203.

100. Carmody, "Energetic consequences of thermal and nonthermal food processing."

101. Roy, "11.5: Neutralization of Fatty Acids and Hydrolysis of Triglycerides."

102. Windey, K., De Preter, V., et al. 2012. "Relevance of protein fermentation to gut health." Mol Nutr Food Res 56(1):184-96.

103. Hoyles, L., and Wallace, R.J. 2010. "Gastrointestinal Tract: Fat Metabolism in the Colon." In Timmis, K.N. (eds) Handbook of Hydrocarbon and Lipid Microbiology. Berlin and Heidelberg, Germany: Springer.

104. Hannou, S.A., Haslam, D.E., et al. 2018. "Fructose metabolism and metabolic disease." J Clin Invest 128(2):545-555.

105. Matthews, M. 2021. "What foods have a high thermic effect?" Accessed on March 21, 2025. https://legionathletics.com/high-thermic-foods/?srsltid=AfmBOorHCNjyUAqB_n-09UTyPqSaQ-1CYgRgIFrycddrpK4FPKyat_JSz.

106. Carmody, "Energetic consequences of thermal and nonthermal food processing."

107. Warneken, F., and Rosati, A.G. 2015. "Cognitive capacities for cooking in chimpanzees." Proc Biol Sc 282(1809):20150229.

108. Michelson, M. 2015. "Chimpanzee Chefs." Accessed March 21, 2025. https://www.calacademy.org/explore-science/chimpanzee-chefs.

109. Goodman, D. "Sprouting 101: What is It and Why Is It Good for You?" Accessed March 21, 2025. https://lynnecohen-foundation.org/the-seam/stories/sprouting-101-what-is-it-and-why-is-it-good-for-you.

110. Knez, E., Kadac-Czapska, K., et al. 2023. "Effect of Fermentation on the Nutritional Quality of the Selected Vegetables and Legumes and Their Health Effects." Life (Basel)13(3): 655.

111. Leeuwendall, N.K., Stanton, C., et al. 2022. "Fermented Foods, Health and the Gut Microbiome." Nutrients 14(7): 1527.

112. Luscombe, N.D., Clifton, P.M., et al. 2003. "Effect of a high-pro-

tein, energy-restricted diet on weight loss and energy expenditure after weight stabilization in hyperinsulinemic subjects." Int J Obes Relat Metab Disord 27(5):582-90.

113. Bax, M.L., Buffière, C., et al. 2013. "Effects of meat cooking, and of ingested amount, on protein digestion speed and entry of residual proteins into the colon: a study in minipigs." PLoS One 8(4):e61252.

114. Et-Chem. 2024. "What Temp Does Collagen Break Down? Cooking Science." Accessed March 21, 2025. https://et-chem.com/what-temp-does-collagen-break-down-cooking-science/.

115. Lee, S., Choi, Y., et al. 2017. "Effect of different cooking methods on the content of vitamins and true retention in selected vegetables." Food Sci Biotechnol 27(2):333-342.

116. Jimenez, P.S., Bangar, S.P., et al. 2023. "Understanding retort processing: A review." Food Sci Nutr 12(3):1545-1563.

117. Harvard T.H. Chan School of Public Health. 2022. "Are Anti-Nutrients Harmful?" Accessed March 21, 2025. https://nutritionsource.hsph.harvard.edu/anti-nutrients/.

118. Wikipedia. 2025. "Phytic acid." Accessed September 4, 2025. https://en.wikipedia.org/wiki/Phytic_acid.

119. World Health Organization. 2023. "Natural toxins in food." Accessed March 21, 2025. https://www.who.int/news-room/fact-sheets/detail/natural-toxins-in-food.

120. Wikipedia. "Secondary metabolites." Accessed September 3, 2025. https://en.wikipedia.org/wiki/Secondary_metabolite.

121. Friday, O.A. 2019. "Plant Toxins." Am J Biomed Sci & Res 4(3).

122. Shewry, P.R., and Halford, N.G. 2002. "Cereal seed storage proteins: structures, properties and role in grain utilization." J Exper Botany 53(370): 947–958,

123. Hariram, H.N., and Park, S.W. 2014. "Edible berries: Bioactive components and their effect on human health." Nutrition 30(2): 134-144.

124. Lambert, J.E. 2013. "Evolutionary Biology of Ape and Monkey Feeding and Nutrition." In Henke, W. Tattersall, I. (eds) Handbook of Paleoanthropology. Berlin and Heidelberg, Germany: Springer.

125. Abbo, S., Gopher, A., et al. 2022. "The Difference between Wild and Domesticated Plants." In Plant Domestication and the Origins of Agriculture in the Ancient Near East. Cambridge, England: Cambridge University Press.

126. Arnarson, A. 2023. "How to Reduce Antinutrients in Food." Accessed March 21, 2025. https://www.healthline.com/nutrition/how-to-reduce-antinutrients.

127. López-Moreno, M., Garcés-Rimón, M., et al. 2022. "Antinutrients: Lectins, goitrogens, phytates and oxalates, friends or foe?" J Funct Foods (89): 104.

128. Ferraro, R., Lillioja, S., et al. 1992. "Lower sedentary metabolic rate in women compared with men." J Clin Invest 90(3):780-4.

129. Palmer, A.K., and Jensen, M.D. 2022. "Metabolic changes in aging humans: current evidence and therapeutic strategies." J Clin Invest 132(16):e158451.

130. Froehle, A.W. 2008. "Climate variables as predictors of basal metabolic rate: new equations." Am J Hum Biol (5):510-29.

131. Wikipedia. 2025. "Skeletal muscle." Accessed September 3, 2025. https://en.wikipedia.org/wiki/Skeletal_muscle.

132. Grattan, B.J., and Connolly-Schoonen, J. 2012. "Addressing weight loss recidivism: a clinical focus on metabolic rate and the psychological aspects of obesity." ISRN Obes 2012:567530.

133. Whitman, S. 2003. "The truth about metabolism: will a slow metabolism make you fat?" Accessed March 22, 2025. https://web.archive.org/web/20110423092125/http://findarticles.com/p/articles/mi_m0846/is_1_23/ai_107488078/?tag=content;col1.

134. Stiegler, P., and Cunliffe, A. 2006. "The Role of Diet and Exercise for the Maintenance of Fat-Free Mass and Resting Metabolic Rate During Weight Loss." Sports Med 36, 239–262.

135. Cleveland Clinic. 2023. "What Is EPOC? (And Why It Matters)." Accessed March 22, 2025. https://health.clevelandclinic.org/understanding-epoc.

136. Johnstone, A.M., Murison S.D., et al. 2005. "Factors influencing variation in basal metabolic rate include fat-free mass, fat mass, age, and circulating thyroxine but not sex, circulating leptin, or triiodothyronine." Am J Clin Nutr 82(5): 941-948.

137. Speakman, J.R., de Jong, J.M.A., et al. 2023. "Total daily energy expenditure has declined over the past three decades due to declining basal expenditure, not reduced activity expenditure." Nat Metab 5, 579–588.

138. Thurber, C., Dugas, L.R., et al. 2019. "Extreme events reveal an alimentary limit on sustained maximal human energy expenditure." Sci Adv 5(6).

139. Dugas, L.R., Harders, R., et al. 2011. "Energy expenditure in adults living in developing compared with industrialized countries: a meta-analysis of doubly labeled water studies." Am J Clin Nutr 93(2):427-41.

140. McAuley, D. 2021. "Schofield equation Basal metabolic rate." Accessed March 24, 2025. https://globalrph.com/medcalcs/schofield-equation-bmr/.

141. Zajac, A., and Mucha, M. 2024. "Harris-Benedict Calculator (Basal Metabolic Rate)." Accessed March 24, 2025. https://www.omnicalculator.com/health/bmr-harris-benedict-equation.

142. Abreu, M. 2023. "Mifflin-St. Jeor for nutrition professionals." Accessed March 24, 2025. https://nutrium.com/blog/mifflin-st-jeor-for-nutrition-professionals/.

143. Gerrior, S., Juan, W., et al. 2006. "An easy approach to calculating estimated energy requirements." Prev Chronic Dis 3(4).

144. Zajac, A., and Mucha, M. 2024. "Katch-McArdle Calculator." Accessed March 24, 2025. https://www.omnicalculator.com/health/bmr-katch-mcardle.

145. Carbone, J.W., and Pasiakos, S.M. 2019. "Dietary Protein and

Muscle Mass: Translating Science to Application and Health Benefit." Nutrients 11(5):1136.

146. Oregon State University. 2025. "24.4 Protein Metabolism." Accessed September 3, 2025. https://open.oregonstate.education/anatomy2e/chapter/protein-metabolism/.

147. Cordain, L., Eaton, S.B., et al. 2005. "Origins and Evolution of the Western Diet: health implications for the 21st century." Am J Clin Nutr 81 (2): 341-354.

148. Lieberman, H.R., Fulgoni, V.L., et al. 2020. "Protein intake is more stable than carbohydrate or fat intake across various US demographic groups and international populations." Am J Clin Nutr 112(1):180-186.

149. Wu, G. 2016. "Dietary protein intake and human health." Food Funct 7(3):1251-65.

150. Pupim, L.B., Martin, C.J., et al. 2013. "Chapter 10 – Assessment of Protein and Energy Nutritional Status." In Nutritional Management of Renal Disease (Third Edition) Eds. Kopple, J.D. et al. San Diego, California: Academic Press.

151. Rand, W.M., Pellett, P.L., et al. 2003. "Meta-analysis of nitrogen balance studies for estimating protein requirements in healthy adults." Am J Clin Nutri 77(1): 109-127.

152. Layman, D.K. 2009. "Dietary Guidelines should reflect new understandings about adult protein needs." Nutr Metab (Lond) 6:12.

153. Layman, "Dietary Guidelines should reflect new understandings about adult protein needs."

154. Dickerson, R.N. 2016. "Nitrogen Balance and Protein Requirements for Critically Ill Older Patients." Nutrients 8(4):226.

155. Delimaris, I. 2013. "Adverse Effects Associated with Protein Intake above the Recommended Dietary Allowance for Adults." ISRN Nutr Jul 18;2013:126929.

156. Delimaris, "Adverse Effects Associated with Protein Intake."

157. Lieberman, "Protein intake is more stable than carbohydrate or fat intake."

158. St. John, A. 2025. "How Much Protein Do Athletes Really Need." Accessed March 25, 2025. https://www.scienceforsport.com/how-much-protein-do-athletes-really-need/.

159. St. John, "How Much Protein Do Athletes Really Need."

160. Examine. 2025. "Optimal Protein Intake Guide & Calculator." Accessed March 24, 2025. https://examine.com/guides/protein-intake/.

161. Examine, "Optimal Protein Intake Guide & Calculator."

162. Hernández-Alonso, P., Salas-Salvadó, J., et al. 2016. 'High dietary protein intake is associated with an increased body weight and total death risk." Clin Nutr.35(2):496-506.

163. Wikipedia. 2025. "Protein digestibility corrected amino acid score." Accessed September 3, 2025. https://en.wikipedia.org/wiki/Protein_digestibility_corrected_amino_acid_score.

164. Cordain, L., Brand Miller, J., et al. 2000. "Plant-animal subsistence ratios and macronutrient energy estimations in worldwide hunter-gatherer diets." Am J Clin Nutr 71(3): 682-92.

165. Paul, C., Leser, S., et al. 2019. "Significant Amounts of Functional Collagen Peptides Can Be Incorporated in the Diet While Maintaining Indispensable Amino Acid Balance." Nutrients 11(5):1079.

166. Lucà-Moretti, M. 1998. "A Comparative, Double-blind ,Triple Crossover Net Nitrogen Utilization Study Confirms the Discovery of the Master Amino Acid Pattern." Ann Royal Natl Acad Med of Spain 115(2).

167. Lucà-Moretti, "A Comparative, Double-blind, Triple Crossover Net Nitrogen Utilization Study."

168. Reeds, P.J. 2000. "Dispensable and Indispensable Amino Acids for Humans." J Nutr 130(7): 1835S-1840S.

169. Ramos-Jiménez, A., Hernández-Torres, R.P., et al. 2024. "An Update of the Promise of Glycine Supplementation for Enhancing Physical Performance and Recovery." Sports (Basel) 12(10):265.

170. Millward, D.J. 2004. "Vernon Young and the development of current knowledge in protein and amino acid nutrition." Am J Clin

Nutr 80(5): 1093-1101.

171. Katagiri, M., and Nakamura, M. 2002. "Animals are dependent on preformed alpha-amino nitrogen as an essential nutrient." IUBMB Life 53(2):125-9.

172. Berry, T.H., Becker, D.E., et al. 1962. "The Limiting Amino Acids in Soybean Protein." J Anim Sci 21(3): 558-561.

173. Wikipedia, "Protein digestibility corrected amino acid score."

174. Simonetti, F. 2023. "Foods High in Protein." Accessed March 25, 2025. https://flaviosimonetti.de/biologische-wertigkeit-tabelle-der-besten-eiweiss-quellen.

175. Kaur, L., Mao, S., et al. 2022. "Alternative proteins vs animal proteins: The influence of structure and processing on their gastro-small intestinal digestion." Trends Food Sci Technol 122: 275-286.

176. Nkhata, S.G., Ayua, E., et al. 2018. "Fermentation and germination improve nutritional value of cereals and legumes through activation of endogenous enzymes." Food Sci Nutr 6(8):2446-2458.

177. Paul, C., Leser, S., et al. 2019. "Significant Amounts of Functional Collagen Peptides Can Be Incorporated in the Diet While Maintaining Indispensable Amino Acid Balance." Nutrients 11(5):1079.

178. Schaafsma, G. 2000. "The Protein Digestibility–Corrected Amino Acid Score." J Nutr 130(7):1865S - 1867S.

179. Schaafsma, "The Protein Digestibility–Corrected Amino Acid Score."

180. Wolfe, R.R., Rutherfurd, S.M., et al. 2016. "Protein quality as determined by the Digestible Indispensable Amino Acid Score: evaluation of factors underlying the calculation." Nutr Rev 74(9):584-99.

181. Yamamoto, K., Igawa, K., et al. 2014. "Biological safety of fish (tilapia) collagen." Biomed Res Int 2014:630757.

182. RxList. 2025. "Gelatin." Accessed March 25, 2025. https://www.rxlist.com/supplements/gelatin.htm.

183. Daneault, A., Prawitt, J., et al. 2017. "Biological effect of hydrolyzed collagen on bone metabolism." Crit Rev Food Sci Nutr

57(9):1922-1937.

184. Reutersward, A.L., Asp, N.G., et al. 1985. "Protein digestibility of pigskin and bovine tendon in rats." Int J Food Sci Techno 20(6):745-752.

185. Yazaki, M., Ito, Y., et al. 2017. "Oral Ingestion of Collagen Hydrolysate Leads to the Transportation of Highly Concentrated Gly-Pro-Hyp and Its Hydrolyzed Form of Pro-Hyp into the Bloodstream and Skin" J Agric Food Chem 265(11):2315-2322

186. Ohara, H., Matsumoto, H., et al. 2007. "Comparison of Quantity and Structures of Hydroxyproline-Containing Peptides in Human Blood after Oral Ingestion of Gelatin Hydrolysates from Different Sources" J Agric Food Chem 55(4):1532-1535

187. Iwai, K., Hasegawa, T., et al. 2005. "Identification of food-derived collagen peptides in human blood after oral ingestion of gelatin hydrolysates." J Agric Food Chem 53(16):6531-6.

188. Laser Reuterswärd, A., Andersson, H., and Asp N.G. 1985. "Digestibility of collagenous fermented sausage in man." Meat Sci 14(2): 105-121,

189. Walrand, S., Chiotelli, E., et al. 2008. "Consumption of a functional fermented milk containing collagen hydrolysate improves the concentration of collagen-specific amino acids in plasma." J Agric Food Chem 56(17):7790-5.

190. Siebecker, A. 2004. Traditional Bone Broth in Modern Health and Disease. Portland, Oregon: NCNM Press.

191. Zague, V., de Freitas, V., et al. 2011. "Collagen hydrolysate intake increases skin collagen expression and suppresses matrix metalloproteinase 2 activity." J Med Food (6):618-24.

192. Shimizu, J., Asami, N., et al. 2015. "Oral collagen-derived dipeptides, prolyl-hydroxyproline and hydroxyprolyl-glycine, ameliorate skin barrier dysfunction and alter gene expression profiles in the skin." Biochem Biophys Res Commun 456(2):626-630.

193. Ohara, H. 2014. "Improvement of extracellular matrix (ECM) in the skin by oral ingestion of collagen hydrolysate." Food Ingredi-

ents J Jpn 219:216-223.

194. Pyun, H.B., Kim, M., et al. 2012. "Effects of Collagen Tripeptide Supplement on Photoaging and Epidermal Skin Barrier in UVB-exposed Hairless Mice." Prev Nutr Food Sci 17(4):245-53.

195. Zhang, Z., Zhao, M., et al. 2011. "Oral administration of skin gelatin isolated from Chum salmon (Oncorhynchus keta) enhances wound healing in diabetic rats." Mar Drugs 9(5):696-711.

196. Shigemura, Y., Iwai, K., et al. 2009. "Effect of Prolyl-hydroxyproline (Pro-Hyp), a food-derived collagen peptide in human blood, on growth of fibroblasts from mouse skin." J Agric Food Chem 57(2):444-9.

197. Ohara, H., Iida, H., et al. 2010. "Effects of Pro-Hyp, a collagen hydrolysate-derived peptide, on hyaluronic acid synthesis using in vitro cultured synovium cells and oral ingestion of collagen hydrolysates in a guinea pig model of osteoarthritis." Biosci Biotechnol Biochem 74(10):2096-9.

198. Ohara, H., Ichikawa, S., et al. 2010. "Collagen-derived dipeptide, proline-hydroxyproline, stimulates cell proliferation and hyaluronic acid synthesis in cultured human dermal fibroblasts." J Dermatol 37(4):330-8.

199. Okawa, T., Yamaguchi, Y., et al. 2012. "Oral administration of collagen tripeptide improves dryness and pruritus in the acetone-induced dry skin model." J Dermatol Sci 66(2):136-43.

200. Proksch, E., Segger, D., et al. 2014. "Oral supplementation of specific collagen peptides has beneficial effects on human skin physiology: a double-blind, placebo-controlled study." Skin Pharmacol Physiol 27(1):47-55.

201. Schunck, M., Zague, V., et al. 2015. "Dietary Supplementation with Specific Collagen Peptides Has a Body Mass Index-Dependent Beneficial Effect on Cellulite Morphology." J Med Food 18(12):1340-8.

202. Inoue, N., Sugihara, F., et al. "Ingestion of bioactive collagen hydrolysates enhance facial skin moisture and elasticity and reduce facial ageing signs in a randomised double-blind placebo-controlled clini-

cal study." J Sci Food Agric 96(12):4077-81.

203. Guillerminet, F., Beaupied, H., et al. 2010. "Hydrolyzed collagen improves bone metabolism and biomechanical parameters in ovariectomized mice: An in vitro and in vivo study." Bone 46(3):827-834.

204. Shaw, G., Lee-Barthel, A., et al. 2017. "Vitamin C-enriched gelatin supplementation before intermittent activity augments collagen synthesis." Am J Clin Nutr 105(1):136-143.

205. Bello, A.E., and Oesser, S. 2006. "Collagen hydrolysate for the treatment of osteoarthritis and other joint disorders: a review of the literature." Curr Med Res Opin 22(11):2221-32.

206. McAlindon, T.E., Nuite, M., et al. 2011. "Change in knee osteoarthritis cartilage detected by delayed gadolinium enhanced magnetic resonance imaging following treatment with collagen hydrolysate: a pilot randomized controlled trial." Osteoarthritis Cartilage 19(4):399-405.

207. Daneault, A., Prawitt, J., et al. 2017. "Biological effect of hydrolyzed collagen on bone metabolism." Crit Rev Food Sci Nutr 57(9):1922-1937.

208. Clark, K.L., Sebastianelli, W., et al. 2008. "24-Week study on the use of collagen hydrolysate as a dietary supplement in athletes with activity-related joint pain." Curr Med Res Opin 24(5):1485-96.

209. Moskowitz, R.W. 2000. "Role of collagen hydrolysate in bone and joint disease." Seminars in Arthritis and Rheumatism 30(2):87-89.

210. Bello, "Collagen hydrolysate for the treatment of osteoarthritis and other joint disorders."

211. Schauss, A.G., Stenehjem, J., et al. 2012. "Effect of the novel low molecular weight hydrolyzed chicken sternal cartilage extract, BioCell Collagen, on improving osteoarthritis-related symptoms: a randomized, double-blind, placebo-controlled trial." J Agric Food Chem 60(16):4096-101.

212. Zdzieblik, D., Oesser, S., et al. 2017. "Improvement of activity-related knee joint discomfort following supplementation of specific collagen peptides." Appl Physiol Nutr Metab 42(6):588-595.

213. Praet, S.F.E., Purdam, C.R., et al. 2019. "Oral Supplementation of Specific Collagen Peptides Combined with Calf-Strengthening Exercises Enhances Function and Reduces Pain in Achilles Tendinopathy Patients." Nutrients 11(1):76.

214. Dressler, P., Gehring, D., et al. 2018. "Improvement of functional ankle properties following supplementation with specific collagen peptides in athletes with chronic ankle instability." Sports Sci Med 17:198-304.

215. Kirmse, M., Oertzen-Hagemann, V., et al. 2019. "Prolonged Collagen Peptide Supplementation and Resistance Exercise Training Affects Body Composition in Recreationally Active Men." Nutrients 11(5):1154.

216. Zdzieblik, D., Oesser, S., et al. 2015. "Collagen peptide supplementation in combination with resistance training improves body composition and increases muscle strength in elderly sarcopenic men: a randomised controlled trial." Br J Nutr 114(8):1237-45.

217. Jendricke, P., Centner, C., et al. 2019. "Specific Collagen Peptides in Combination with Resistance Training Improve Body Composition and Regional Muscle Strength in Premenopausal Women: A Randomized Controlled Trial." Nutrients 11(4):892.

218. Tang, L., Sakai, Y., et al. 2015. "Effects of oral administration of tripeptides derived from type I collagen (collagen tripeptide) on atherosclerosis development in hypercholesterolemic rabbits." J Biosci Bioeng 119(5):558-63.

219. Zhang, Y., Kouguchi, T., et al. 2010. "Chicken collagen hydrolysate protects rats from hypertension and cardiovascular damage." J Med Food 13(2):399-405.

220. Saito, M., Kiyose, C., et al. 2009. "Effect of collagen hydrolysates from salmon and trout skins on the lipid profile in rats." J Agric Food Chem 57(21):10477-82.

221. Iba, Y., Yokoi, K., et al. 2016. "Oral Administration of Collagen Hydrolysates Improves Glucose Tolerance in Normal Mice Through GLP-1-Dependent and GLP-1-Independent Mechanisms." J Med Food 19(9):836-43.

222. Wang, X., Yu, H., et al. 2018. "Effect and mechanism of oyster hydrolytic peptides on spatial learning and memory in mice." RSC Adv 8:6125–6135.

223. Pei, X., Yang, R., et al. 2010. "Marine collagen peptide isolated from Chum Salmon (Oncorhynchus keta) skin facilitates learning and memory in aged C57BL/6J mice." Food Chem 118(2):333-340.

224. Ucar, B., and Humpel, C. 2018. "Collagen for brain repair: therapeutic perspectives." Neural Regen Res 13(4):595-598.

225. Koizumi, S., Inoue, N., et al. 2019. "Effects of Collagen Hydrolysates on Human Brain Structure and Cognitive Function: A Pilot Clinical Study." Nutrients 12(1):50.

226. Koizumi, "Effects of Collagen Hydrolysates on Human Brain Structure and Cognitive Function."

227. Gregorio, I., Braghetta, P., et al. 2018. "Collagen VI in healthy and diseased nervous system." Dis Model Mech 11(6):32946.

228. Bensky, D., Gamble, A., et al. 1993. Chinese Herbal Medicine Materia Medica. Seattle, Washington: Eastland Press.

229. Graham, M.F., Drucker, D.E., et al. 1987. "Collagen synthesis by human intestinal smooth muscle cells in culture." Gastroenterology 92(2):400-5.

230. Liang, Q., Wang, L., et al. 2014. "Hydrolysis kinetics and antioxidant activity of collagen under simulated gastrointestinal digestion." J Funct Foods 11:493-499.

231. Rogov, I.A., Kovalev, Y.I., et al. 1992. "Collagen and its rational content in meat products 2: Experiments with growing rats." Meat Sci 31(2):147-53.

232. Rogov, I.A., Tokaev, E.S., et al. 1992. "Collagen and its rational content in meat products: Part 1. Analytical studies." Meat Sci 31(1):35-42.

233. Paul, C., Leser, S., et al. 2019. "Significant Amounts of Functional Collagen Peptides Can Be Incorporated in the Diet While Maintaining Indispensable Amino Acid Balance." Nutrients 11(5):1079.

234. Deane, C.S., Bass, J.J., et al. 2020. "Animal, Plant, Collagen and

Blended Dietary Proteins: Effects on Musculoskeletal Outcomes." Nutrients 12(9):2670.

235. Hays, N.P., Kim, H., et al. 2009. "Effects of whey and fortified collagen hydrolysate protein supplements on nitrogen balance and body composition in older women." J Am Diet Assoc 109(6):1082-7.

236. Wasserman, D.H. 2009. "Four g of glucose." Am J Physiol Endocrinol Metab 296(1):E11-21.

237. Kapogiannis, D., and Avgerinos, K.I. 2020. "Chapter Three – Brain glucose and ketone utilization in brain aging and neurodegenerative diseases" Int Rev Neurobio 154:79-110.

238. El Bacha, T., Luz, M., et al. 2010. "Dynamic Adaptation of Nutrient Utilization in Humans." Nature Education 3(9):8.

239. Pietrangelo, A. 2022. "How Much of Our Brain Do We Use? And Other Questions Answered." Accessed March 26, 2025. https://www.healthline.com/health/how-much-of-our-brain-do-we-use.

240. Gillespie, K.M., White, M.J., et al. 2023. "The Impact of Free and Added Sugars on Cognitive Function: A Systematic Review and Meta-Analysis." Nutrients 16(1):75.

241. The Cleveland Clinic. 2025. "Blood Glucose (Sugar) Test." Accessed March 26, 2025. https://my.clevelandclinic.org/health/diagnostics/12363-blood-glucose-test.

242. Seery, C. 2022. "Living with Diabetes: Blood Sugar Level Range." Accessed March 26, 2025. https://www.diabetes.co.uk/diabetes_care/blood-sugar-level-ranges.html#google_vignette.

243. Ludwig, D.S., Hu, F.B., et al. 2018. "Dietary carbohydrates: role of quality and quantity in chronic disease." BMJ 361:k2340.

244. Ludwig, D.S., Majzoub, J.A., et al. 1999. "High glycemic index foods, overeating, and obesity." Pediatrics 103(3):E26.

245. Roberts, S.B. 2003. "Glycemic index and satiety." Nutr Clin Care 6(1):20-6.

246. Anderson, G.H., and Woodend, D. 2003. "Effect of glycemic carbohydrates on short-term satiety and food intake." Nutr Rev 61(5 Pt 2):S17-26.

247. Mohanty, P., Ghanim, H., et al. 2002. "Both lipid and protein intakes stimulate increased generation of reactive oxygen species by polymorphonuclear leukocytes and mononuclear cells." Am J Clin Nutr 75(4):767-72.

248. Morelli, R., Russo-Volpe, S., et al. 2003. "Fenton-dependent damage to carbohydrates: free radical scavenging activity of some simple sugars." J Agric Food Chem 51(25):7418-25.

249. Prasad, K., and Dhar, I. 2014. "Oxidative stress as a mechanism of added sugar-induced cardiovascular disease." Int J Angiol 23(4):217-26.

250. Kawano, Y., Edwards, M., et al. 2022. "Microbiota imbalance induced by dietary sugar disrupts immune-mediated protection from metabolic syndrome." Cell 185(19):3501-3519.

251. Yang, Q., Zhang, Z., et al. 2014. "Added sugar intake and cardiovascular diseases mortality among US adults." JAMA Intern Med 174(4):516-24.

252. Ludwig, D.S., Hu, F.B., et al. 2018. "Dietary carbohydrates: role of quality and quantity in chronic disease." BMJ 361:k2340.

253. Sharma, R., and Tandon, R.K. 2013. "Nutrition, Dietary Fibers, and Cholelithiasis: Apple Pulp, Fibers, Clinical Trials." In Bioactive Food as Dietary Interventions for Liver and Gastrointestinal Disease. Amsterdam, The Netherlands: Elsevier.

254. Schulze, M.B., Manson, J.E., et al. 2004. "Sugar-sweetened beverages, weight gain, and incidence of type 2 diabetes in young and middle-aged women." JAMA 292(8):927-34

255. Fung, T.T., Malik, V., et al. 2009. "Sweetened beverage consumption and risk of coronary heart disease in women." Am J Clin Nutr 89(4):1037-42.

256. Ludwig, "Dietary carbohydrates: role of quality and quantity in chronic disease."

257. Swaminathan, S., Dehghan, M., et al. 2021. "Associations of cereal grains intake with cardiovascular disease and mortality across 21 countries in Prospective Urban and Rural Epidemiology study: prospective cohort study." BMJ 372:m4948.

258. Swaminathan, "Associations of cereal grains intake with cardiovascular disease and mortality."

259. Ludwig, "Dietary carbohydrates: role of quality and quantity in chronic disease."

260. Hu, E.A., Pan, A., et al. 2012. "White rice consumption and risk of type 2 diabetes: meta-analysis and systematic review." BMJ 344:e1454.

261. Ludwig, "Dietary carbohydrates: role of quality and quantity in chronic disease."

262. Sreeraman, V.R. 2008. "White Bread Breakfast Raises Diabetes, Heart Disease and Cancer Risk." Accessed March 27, 2025. https://www.medindia.net/news/white-bread-breakfast-raises-diabetes-heart-disease-and-cancer-risk-33968-1.htm.

263. Halkjaer, J., Tjønneland, A., et al. 2009. "Dietary predictors of 5-year changes in waist circumference." J Am Diet Assoc 109(8):1356-66.

264. Pharr, J.R. 2010. "Carbohydrate Consumption and Fatigue: A Review." NJPH 7(1):Article 6.

265. Hawkins, M.A.W., Keirns, N.G., et al. 2018. "Carbohydrates and cognitive function." Curr Opin Clin Nutr Metab Care 21(4):302-307.

266. Rahbar, S., and Figarola, J.L. 2003. "Novel inhibitors of advanced glycation endproducts." Arch Biochem Biophys 419(1):63-79.

267. Seetharaman, S. 2016. "The Influences of Dietary Sugar and Related Metabolic Disorders on Cognitive Aging and Dementia." In Molecular Basis of Nutrition and Aging Amsterdam, The Netherlands: Elsevier.

268. Ludwig, "Dietary carbohydrates: role of quality and quantity in chronic disease."

269. Rippe, J.M., and Angelopoulos, T.J. 2016. "Relationship between Added Sugars Consumption and Chronic Disease Risk Factors: Current Understanding." Nutrients 8(11):697.

270. Chiu, C.J., and Taylor, A. 2011. "Dietary hyperglycemia, gly-

cemic index and metabolic retinal diseases." Prog Retin Eye Res 30(1):18-53.

271. Jenkins, D.J.A,. Dehghan, M., et al. 2021. "Glycemic Index, Glycemic Load, and Cardiovascular Disease and Mortality." N Engl J Med 384(14):1312-1322.

272. O'Riordan, M. 2021. "High Glycemic Index Diets Linked to CVD and Mortality." Accessed March 27, 2025. https://www.tctmd.com/news/high-glycemic-index-diets-linked-cvd-and-mortality.

273. Fan, J., Song, Y., et al. 2012. "Dietary glycemic index, glycemic load, and risk of coronary heart disease, stroke, and stroke mortality: a systematic review with meta-analysis." PLoS One 7(12):e52182.

274. Augustin, L.S.A., Kendall, C.W.C., et al. 2015. "Glycemic index, glycemic load and glycemic response: An International Scientific Consensus Summit from the International Carbohydrate Quality Consortium (ICQC)." Nutr Metab Cardiovasc Dis 25(9):795-815.

275. Beulens, J.W., de Bruijne, L.M., et al. 2007. "High dietary glycemic load and glycemic index increase risk of cardiovascular disease among middle-aged women: a population-based follow-up study." J Am Coll Cardiol 50(1):14-21.

276. Zafar, M.I., Mills, K.E., et al. 2019. "Low-glycemic index diets as an intervention for diabetes: a systematic review and meta-analysis." Am J Clin Nutr 110(4):891-902.

277. Ludwig, "Dietary carbohydrates: role of quality and quantity in chronic disease."

278. Vega-López, S., Venn, B.J., et al. 2018. "Relevance of the Glycemic Index and Glycemic Load for Body Weight, Diabetes, and Cardiovascular Disease." Nutrients 10(10):1361.

279. Shahdadian, F., Saneei, P., et al. 2019. "Dietary glycemic index, glycemic load, and risk of mortality from all causes and cardiovascular diseases: a systematic review and dose-response meta-analysis of prospective cohort studies." Am J Clin Nutr 110(4):921-937.

280. Willcox, B.J., Willcox, D.C., et al. 2007. "Caloric restriction, the traditional Okinawan diet, and healthy aging: the diet of the world's

longest-lived people and its potential impact on morbidity and life span." Ann NY Acad Sci 1114: 434–55.

281. Willcox, D.C., Willcox, B.J., et al. 2009. "The Okinawan Diet: Health Implications of a Low-Calorie, Nutrient-Dense, Antioxidant-Rich Dietary Pattern Low in Glycemic Load." J Amer Coll Nutri 28:500S-516S.

282. Sho, H. 2001. "History and characteristics of Okinawan longevity food." Asia Pacific J Clin Nutr 10(2):159-164.

283. Kaplan. H., Thompson, R.C., et al. 2017. "Coronary atherosclerosis in indigenous South American Tsimane: a cross-sectional cohort study." Lancet 389(10080):1730-1739.

284. Pritikin, N. 1983. The Pritikin Promise: 28 Days to a Longer, Healthier Life. New York, New York: Simon & Schuster.

285. Ornish, D. 2025. "Over 45 Years Pioneering Research." Accessed January 27, 2025. https://pmri.org/.

286. Wenner Moyer, M., and Ornish, D. 2015. "Why Almost Everything Dean Ornish Says about Nutrition Is Wrong. UPDATED: With Dean Ornish's Response." Accessed March 27, 2025. https://www.scientificamerican.com/article/why-almost-everything-dean-ornish-says-about-nutrition-is-wrong/.

287. Gunnars, K., and Northrop, A. 2023. "The Atkins Diet: Everything You Need to Know." Accessed March 27, 2025. https://www.healthline.com/nutrition/atkins-diet-101.

288. Lin, P.J., and Borer, K.T. 2016. "Third Exposure to a Reduced Carbohydrate Meal Lowers Evening Postprandial Insulin and GIP Responses and HOMA-IR Estimate of Insulin Resistance." PLoS One 11(10):e0165378.

289. Roberts, M.N., Wallace, M.A., et al. "A Ketogenic Diet Extends Longevity and Healthspan in Adult Mice." Cell Metab 26(3):539-546.

290. Martin McGill, K.J., Jackson, C.F., et al. 2018. "Ketogenic diets for drug resistant epilepsy." Cochrane Database of Systematic Reviews 11: Article #CD001903.

291. Kossoff, E.H., Zupec-Kania, B.A., et al. 2009. "Ketogenic diets: an

update for child neurologists." J Child Neurol 24(8):979–988.

292. Hashimoto, Y., Fukuda, T., et al. 2016. "Impact of low-carbohydrate diet on body composition: meta-analysis of randomized controlled studies." Obesity Reviews 17(6):499–509.

293. Nonas, C.A., and Dolins, K.R. 2012. "Dietary intervention approaches to the treatment of obesity." In Textbook of Obesity Hoboken, New Jersey: John Wiley & Sons.

294. Dashti, H.M., Mathew, T.C., et al. "Long-term effects of a ketogenic diet in obese patients." Exp Clin Cardiol 9(3):200-5.

295. Kossoff, E.H., and Wang, H.S. 2013. "Dietary therapies for epilepsy." Machine Biomed J 36(1):2-8.

296. Turner, Z., and Kossoff, E.H. 2006. "The ketogenic and Atkins diets: recipes for seizure control." Pract Gastroenterol 29(6):53, 56, 58, 61–62, 64.

297. Bergqvist, A.G. 2011. "Long-term monitoring of the ketogenic diet: Do's and Don'ts." Epilepsy Res 100(3):261-266.

298. Sussman, D., van Eede, M., et al. 2013. "Effects of a ketogenic diet during pregnancy on embryonic growth in the mouse." BMC Pregnancy Childbirth 8;13:109.

299. Foster, G.D., Wyatt, H.R., et al. 2003. "A randomized trial of a low-carbohydrate diet for obesity." N Engl J Med 348(21):2082-90.

300. Gunnars, K. 2020. "23 Studies on Low Carb and Low Fat Diets – Time to Retire the Fad." Accessed March 27, 2025. https://www.healthline.com/nutrition/23-studies-on-low-carb-and-low-fat-diets.

301. Hall, K.D. Guo, J. et al. 2021. "Effect of a plant-based, low-fat diet versus an animal-based, ketogenic diet on ad libitum energy intake." Nat Med 27:344–353.

302. Miller, M., Beach, V., et al. 2009. "Comparative effects of three popular diets on lipids, endothelial function, and C-reactive protein during weight maintenance." J Am Diet Assoc 109(4):713-7.

303. Harvard T.H. Chan School of Public Health. "PURE study makes

headlines, but the conclusions are misleading." 2017. Accessed March 27, 2025. https://nutritionsource.hsph.harvard.edu/2017/09/08/pure-study-makes-headlines-but-the-conclusions-are-misleading/.

304. Seidelmann, S.B., Claggett, B., et al. 2018. "Dietary carbohydrate intake and mortality: a prospective cohort study and meta-analysis." Lancet Public Health 3(9):e419-e428.

305. Matthews-King, A. 2018. "Low carb diets shorten your life unless you are mostly vegetarian, study suggests." Accessed March 27, 2025. https://www.independent.co.uk/news/health/low-carb-diet-protein-lean-meat-dairy-weight-loss-keto-fat-vegetarian-wholegrain-lancet-a8494621.html.

306. Dehghan, M., Mente, A., et al. 2017. "Associations of fats and carbohydrate intake with cardiovascular disease and mortality in 18 countries from five continents (PURE): a prospective cohort study." Lancet 390(10107):2050-2062.

307. Laughlin, M.R. 2014. "Normal roles for dietary fructose in carbohydrate metabolism." Nutrients 6(8):3117-29.

308. Stephan, B.C., Wells, J.C., et al. 2010. "Increased fructose intake as a risk factor for dementia." J Gerontol A Biol Sci Med Sci 65(8):809-14.

309. Ludwig, "Dietary carbohydrates: role of quality and quantity in chronic disease."

310. Pereira, R.M., Botezelli, J.D., et al. 2017. "Fructose Consumption in the Development of Obesity and the Effects of Different Protocols of Physical Exercise on the Hepatic Metabolism." Nutrients 9(4):405.

311. Laughlin, "Normal roles for dietary fructose in carbohydrate metabolism."

312. Ludwig, "Dietary carbohydrates: role of quality and quantity in chronic disease."

313. Ludwig, D.S. 2013. "Examining the Health Effects of Fructose." JAMA 310(1):33-34.

314. Sunehag, A., Tigas, S., et al. 2003. "Contribution of plasma galac-

tose and glucose to milk lactose synthesis during galactose ingestion." J Clin Endocrinol Metab 88(1):225–9.

315. Parada Venegas, D., De la Fuente, M.K., et al. 2019. "Short Chain Fatty Acids (SCFAs) – Mediated Gut Epithelial and Immune Regulation and Its Relevance for Inflammatory Bowel Diseases." Front Immunol 10:277.

316. Wong, J.M., de Souza, R., et al. 2006. "Colonic health: fermentation and short chain fatty acids." J Clin Gastroenterol 40(3):235-43.

317. Stilling, R.M., van de Wouw, M., et al. 2016. "The neuropharmacology of butyrate: The bread and butter of the microbiota-gut-brain axis?" Neurochem Int 99:110-132.

318. Moszak, M., Szuli ska, M., et al. 2020. "You Are What You Eat – The Relationship between Diet, Microbiota, and Metabolic Disorders – A Review." Nutrients 12(4):1096.

319. The Cleveland Clinic. 2025. "The Gut Microbiome." Accessed March 27, 2025. https://my.clevelandclinic.org/health/body/25201-gut-microbiome.

320. Parada Venegas, D., De la Fuente, M.K., et al. 2019. "Short Chain Fatty Acids (SCFAs) – Mediated Gut Epithelial and Immune Regulation and Its Relevance for Inflammatory Bowel Diseases." Front Immunol 11;10:277.

321. Moszak, "You Are What You Eat."

322. Holscher, H.D. 2017. "Dietary fiber and prebiotics and the gastrointestinal microbiota." Gut Microbe 8(2):172-184.

323. Holscher, "Dietary fiber and prebiotics and the gastrointestinal microbiota."

324. Singh, R.K., Chang, H.W., et al. 2017. "Influence of diet on the gut microbiome and implications for human health." J Transl Med 15(1):73.

325. Frame, L.A., Costa, E., et al. 2020. "Current explorations of nutrition and the gut microbiome: a comprehensive evaluation of the review literature." Nutr Rev 78(10):798-812.

326. Moszak, "You Are What You Eat."

327. Frame, "Current explorations of nutrition and the gut microbiome."

328. Ramos Meyers, G., Samouda, H., et al. 2022. "Short Chain Fatty Acid Metabolism in Relation to Gut Microbiota and Genetic Variability." Nutrients 14(24):5361.

329. Salleh, S.N., Fairus, A.A.H., et al. 2019. "Unravelling the Effects of Soluble Dietary Fibre Supplementation on Energy Intake and Perceived Satiety in Healthy Adults: Evidence from Systematic Review and Meta-Analysis of Randomised-Controlled Trials." Foods 8(1):15.

330. Ramezani, F., Pourghazi, F., et al. 2024. "Dietary fiber intake and all-cause and cause-specific mortality: An updated systematic review and meta-analysis of prospective cohort studies." Clin Nutr 43(1):65-83.

331. Fu, L., Zhang, G., et al. 2022. "Associations between dietary fiber intake and cardiovascular risk factors: An umbrella review of meta-analyses of randomized controlled trials." Front Nutr 12(9):972399.

332. Wikipedia. 2025. "Diet and Cancer." Accessed September 4, 2025. https://en.wikipedia.org/wiki/Diet_and_cancer.

333. Chen, J.P., Chen, G.C., et al. 2018. "Dietary Fiber and Metabolic Syndrome: A Meta-Analysis and Review of Related Mechanisms." Nutrients 10:24.

334. Kabisch, S., Hajir, J., et al. 2025. "Impact of Dietary Fiber on Inflammation in Humans." Int J Mol Sci 26: 2000.

335. Anderson, J.W., Baird, P., et al. 2009. "Health benefits of dietary fiber." Nutr Rev 67(4):188-205.

336. Lattimer, J.M., and Haub, M.D. 2010. "Effects of dietary fiber and its components on metabolic health." Nutrients 2(12):1266-89.

337. Anderson, J.W., Randles, K.M., et al. 2004. "Carbohydrate and fiber recommendations for individuals with diabetes: a quantitative assessment and meta-analysis of the evidence." J Am Coll Nutr 23(1):5-17.

338. Frame, "Current explorations of nutrition and the gut microbiome."

339. Roy, C.C., Kien, C.L., et al. 2006. "Short-chain fatty acids: ready for prime time?" Nutri Clin Pract 21(4):351–66.

340. Carabotti, M., Scirocco, A., et al. 2015. "The gut-brain axis: interactions between enteric microbiota, central and enteric nervous systems." Ann Gastroenterol 28(2):203-209.

341. Cryan, J.F., O'Riordan, K.J., et al. 2019. "The Microbiota-Gut-Brain Axis." Physiological Reviews 99(4):1877-2013.

342. Cowan, C.S.M., Hoban, A.E., et al. 2018. "Gutsy Moves: The Amygdala as a Critical Node in Microbiota to Brain Signaling." Bioessays 40(1).

343. Saxena, R., and Sharma, V.K. 2016. "A Metagenomic Insight Into the Human Microbiome: Its Implications in Health and Disease." In Medical and Health Genomics Amsterdam, The Netherlands; Elsevier Science.

344. Carmel, J., Ghanayem, N., et al. 2023. "Bacteroides is increased in an autism cohort and induces autism-relevant behavioral changes in mice in a sex-dependent manner." NPJ Biofilms Microbiomes 9(1):103.

345. Morais, L.H., Golubeva, A.V., et al. 2020. "Enduring Behavioral Effects Induced by Birth by Caesarean Section in the Mouse." Curr Biol 30(19):3761-3774.

346. Johnson, K.V. 2020. "Gut microbiome composition and diversity are related to human personality traits." Hum Microb J 15.

347. De Palma, G., Lynch, M.D., et al. 2017. "Transplantation of fecal microbiota from patients with irritable bowel syndrome alters gut function and behavior in recipient mice." Sci Transl Med 9(379).

348. Yano, J.M., Yu, K., et al. 2015. "Indigenous bacteria from the gut microbiota regulate host serotonin biosynthesis." Cell 161(2):264-76.

349. Tette, F.M., Kwofie, S.K., et al. 2022. "Therapeutic Anti-Depressant Potential of Microbial GABA Produced by Lactobacillus rhamnosus Strains for GABAergic Signaling Restoration and Inhibition of Addiction-Induced HPA Axis Hyperactivity." Curr Issues Mol Biol

44(4):1434-1451.

350. Cryan, "The Microbiota-Gut-Brain Axis."

351. Chen, Y., Xu, J., et al. 2021. "Regulation of Neurotransmitters by the Gut Microbiota and Effects on Cognition in Neurological Disorders." Nutrients 13(6):2099.

352. Dolan, E.W. 2023. "New study links disturbed energy metabolism in depressed individuals to disruption of the gut microbiome." Accessed March 28, 2025. https://www.psypost.org/new-study-links-disturbed-energy-metabolism-in-depressed-individuals-to-disruption-of-the-gut-microbiome/.

353. Minot, S., Sinha, R., et al. 2011. "The human gut virome: inter-individual variation and dynamic response to diet." Genome Res 21(10):1616-25.

354. Mercola, J., and D'Adamo, C.R. 2023. "Linoleic Acid: A Narrative Review of the Effects of Increased Intake in the Standard American Diet and Associations with Chronic Disease." Nutrients 15: 3129.

355. Werner, A. ,Kuipers, F., et al. 2000-2013. "Fat Absorption and Lipid Metabolism in Cholestasis." In Madame Curie Bioscience Database [Internet]. Austin, Texas: Landes Bioscience.

356. Roopashree, P.G., Shetty, S.S., et al. 2021. "Effect of medium chain fatty acid in human health and disease." J Funct Foods 87:104724.

357. Heart UK. 2018. "Blood Fats Explained." Accessed on March 28, 2025. https://www.heartuk.org.uk/downloads/health-professionals/publications/blood-fats-explained.pdf.

358. Novak, E.M., Dyer, R.A., et al. 2008. "High dietary omega-6 fatty acids contribute to reduced docosahexaenoic acid in the developing brain and inhibit secondary neurite growth." Brain Res 1237:136-45.

359. GERLI. "Animal Fats." Accessed March 28, 2025. https://cyberlipid.gerli.com/description/simple-lipids/triacylglycerols/animal-fats/.

360. Carson, J.A.S., Lichtenstein, A.H., et al. 2019. "Dietary Cholesterol and Cardiovascular Risk: A Science Advisory From the American Heart Association." Circulation 41(3): e39-e53.

361. Soliman, G.A. 2018. "Dietary Cholesterol and the Lack of Evidence in Cardiovascular Disease." Nutrients 10(6):780.

362. Corliss, J. 2024. "Rethinking HDL cholesterol." Accessed March 28, 2025. https://www.health.harvard.edu/heart-health/rethinking-hdl-cholesterol.

363. National Heart, Lung, and Blood Institute. 2022. "Study challenges 'good' cholesterol's role in universally predicting heart disease risk." Accessed on March 28, 2025. https://www.nhlbi.nih.gov/news/2022/study-challenges-good-cholesterols-role-universally-predicting-heart-disease-risk.

364. Krans, B. 2016. "'Bad' Cholesterol May Have a Bad Rap." Accessed March 28, 2025. https://www.healthline.com/health-news/bad-cholesterol-may-have-bad-rap.

365. Scott, P. 2010. "Bad cholesterol: It's not what you think." Accessed on March 28, 2025. https://www.nbcnews.com/health/health-news/bad-cholesterol-it-s-not-what-you-think-flna1c9442109.

366. Superko, H., and Garrett, B. 2022. "Small Dense LDL: Scientific Background, Clinical Relevance, and Recent Evidence Still a Risk Even with 'Normal' LDL-C Levels." Biomedicines 10(4):829.

367. Time Staff. 2014. "Ending the War on Fat." Accessed September 4, 2025. https://time.com/2863227/ending-the-war-on-fat/.

368. Teicholz, N. 2023. "A short history of saturated fat: the making and unmaking of a scientific consensus." Curr Opin Endocrinol Diabetes Obes 30(1):65-71.

369. Fernandez, M.L., and Murillo, A.G. 2022. "Is There a Correlation between Dietary and Blood Cholesterol? Evidence from Epidemiological Data and Clinical Interventions." Nutrients 14(10):2168.

370. Brown, M.S., and Goldstein, J.L. 1997. "The SREBP pathway: regulation of cholesterol metabolism by proteolysis of a membrane-bound transcription factor." Cell 89(3):331–340.

371. Brown, "The SREBP pathway."

372. Gui, Y., Zheng, H., et al. 2022. "Foam Cells in Atherosclerosis: Novel Insights Into Its Origins, Consequences, and Molecular Mechanisms." Front Cardiovasc Med 9:845942.

373. Alfaddagh, A., Martin, S.S., et al. 2020. "Inflammation and cardiovascular disease: From mechanisms to therapeutics." Am J Prev Cardiol 4:100130.

374. Russell, D.W. 2000. "Oxysterol biosynthetic enzymes." Biochimica et Biophysica Acta (BBA) – Molecular and Cell Biology of Lipids 1529(1-3):126-135.

375. Juste, C., and Gérard, P. 2021. "Cholesterol-to-Coprostanol Conversion by the Gut Microbiota: What We Know, Suspect, and Ignore." Microorganisms 9(9):1881.

376. Rye, K.A., Bursill, C.A., et al. 2009. "The metabolism and anti-atherogenic properties of HDL." J. Lipid Res 50:S195–S200.

377. Ras, R.T., Geleijnse, J.M., et al. 2014. "LDL-cholesterol-lowering effect of plant sterols and stanols across different dose ranges: a meta-analysis of randomised controlled studies." Brit J of Nutri 112(2):214–219.

378. DiNicolantonio, J.J., and O'Keefe, J.H. 2018. "Effects of dietary fats on blood lipids: a review of direct comparison trials." Open Heart 5(2):e000871.

379. Estruch, R., Ros, E., et al. 2018. "Primary Prevention of Cardiovascular Disease with a Mediterranean Diet Supplemented with Extra-Virgin Olive Oil or Nuts." N Engl J Med 378(25):e34.

380. Hamley, S. 2017. "The effect of replacing saturated fat with mostly n-6 polyunsaturated fat on coronary heart disease: a meta-analysis of randomised controlled trials." Nutr J 16(1):30.

381. Abdelhamid, A.S., Brown T.J., et al. 2020. "Omega 3 fatty acids for the primary and secondary prevention of cardiovascular disease." Cochrane Database of Systematic Reviews (3):CD003177.

382. Kris-Etherton, P.M., and Yu, S. 1997. "Individual fatty acid effects on plasma lipids and lipoproteins: human studies." Am J Clin Nutr 65(5):1628S-1644S.

383. Katan, M.B., Zock, P.L., et al. 1994. "Effects of fats and fatty acids on blood lipids in humans: an overview." Am J Clin Nutr 60(6):1017S-1022S.

384. National Cholesterol Education Program. 2002. "Detection, Evaluation, and Treatment of High Blood Cholesterol in Adults (Adult Treatment Panel III)." Circulation 106(25):3143-421.

385. American Heart Association. 2002. "II. Rationale for Intervention." Circulation 106(25):3163-3223.

386. American Heart Association, "II. Rationale for Intervention."

387. Lewington, S., Whitlock, G., et al. 2007. "Blood cholesterol and vascular mortality by age, sex, and blood pressure: a meta-analysis of individual data from 61 prospective studies with 55,000 vascular deaths." Lancet 370(9602):1829–1839.

388. American Heart Association, "II. Rationale for Intervention."

389. Yoshida, H., Ito, K., et al. 2021. "Clinical Significance of Intermediate-Density Lipoprotein Cholesterol Determination as a Predictor for Coronary Heart Disease Risk in Middle-Aged Men." Front Cardiovasc Med 8:756057.

390. McMillen, M. 2024. "Cholesterol Particle Size: What It Is and Why It Matters." Accessed March 28, 2025. https://www.healthcentral.com/condition/high-cholesterol/cholesterol-particle-size.

391. National Cholesterol Education Program, "Detection, Evaluation, and Treatment of High Blood Cholesterol."

392. Astrup, A., Magkos, F., et al. 2020. "Saturated Fats and Health: A Reassessment and Proposal for Food-Based Recommendations: JACC State-of-the-Art Review." J Am Coll Cardiol 76(7):844-857.

393. Heileson, J.L. 2020. "Dietary saturated fat and heart disease: a narrative review." Nutr Rev 78(6):474-485.

394. Siri-Tarino, P.W., Sun, Q., et al. 2010. "Meta-analysis of prospective cohort studies evaluating the association of saturated fat with cardiovascular disease." Am J Clin Nutr 91(3):535-46.

395. Schwab, U., Lauritzen, L., et al. 2014. "Effect of the amount and type of dietary fat on cardiometabolic risk factors and risk of developing type 2 diabetes, cardiovascular diseases, and cancer: a sys-

tematic review." Food Nutr Res 10;58.

396. Oguntibeju, O.O., Esterhuyse, A.J., et al. 2009. "Red palm oil: nutritional, physiological and therapeutic roles in improving human wellbeing and quality of life." Br J Biomed Sci 66(4):216-22.

397. Howard, B.V., Lee, E.T., et al. 1999. "Rising tide of cardiovascular disease in American Indians. The Strong Heart Study." Circulation 99(18):2389-95.

398. Sjölander P. 2011. "What is known about the health and living conditions of the indigenous people of northern Scandinavia, the Sami?" Glob Health Action 4.

399. Sanders, R. 2015. "What the Inuit can tell us about omega-3 fats and 'paleo' diets." Accessed March 28, 2025. https://news.berkeley.edu/2015/09/17/what-the-inuit-can-tell-us-about-omega-3-fats-and-paleo-diets/#:~:text=On%20their%20traditional%20diet%2C%20rich,Berkeley%20professor%20of%20integrative%20biology.

400. Berger, M.E., Smesny, S., et al. 2017. "Omega-6 to omega-3 polyunsaturated fatty acid ratio and subsequent mood disorders in young people with at-risk mental states: a 7-year longitudinal study." Transl Psychiatry 7(8): e1220.

401. DiNicolantonio, J.J., O'Keefe, J.H., 2020. "The Importance of Marine Omega-3s for Brain Development and the Prevention and Treatment of Behavior, Mood, and Other Brain Disorders." Nutrients 12(8): 2333.

402. Ross, B.M., Seguin, J., et al. 2007. "Omega-3 fatty acids as treatments for mental illness: which disorder and which fatty acid?" Lipids Health Dis 6: 21.

403. Djuricic, I., and Calder, P.C. 2021 "Beneficial Outcomes of Omega-6 and Omega-3 Polyunsaturated Fatty Acids on Human Health: An Update for 2021." Nutrients 13(7): 2421.

404. LeWine, H.E. 2024. "Fish oil: friend or foe?" Accessed September 4, 2025. https://www.health.harvard.edu/blog/fish-oil-friend-or-foe-201307126467.

405. Simopoulos, A.P. 1991. "Omega-3 fatty acids in health and disease

and in growth and development." Am J Clin Nutr 54(3):438-63.

406. Mariamenatu, A.H., and Abdu, E.M. 2021. "Overconsumption of Omega-6 Polyunsaturated Fatty Acids (PUFAs) versus Deficiency of Omega-3 PUFAs in Modern-Day Diets: The Disturbing Factor for Their 'Balanced Antagonistic Metabolic Functions' in the Human Body." J Lipids 2021:8848161.

407. Marangoni, F., Agostoni, C., et al. 2020. "Dietary linoleic acid and human health: Focus on cardiovascular and cardiometabolic effects." Atherosclerosis 292:90-98.

408. Johnson, G.H., and Fritsche, K. 2012. "Effect of dietary linoleic acid on markers of inflammation in healthy persons: a systematic review of randomized controlled trials." J Acad Nutr Diet 112(7):1029-41.

409. Turpeinen, A.M., Basu, S., et al. 1998. "A high linoleic acid diet increases oxidative stress in vivo and affects nitric oxide metabolism in humans." Prostaglandins Leukot Essent Fatty Acids 59(3):229-33.

410. Ramsden C.E., Ringel A., et al. 2012. "Lowering dietary linoleic acid reduces bioactive oxidized linoleic acid metabolites in humans." Prostaglandins Leukot Essent Fatty Acids 87(4-5):135-41.

411. Jandacek, R.J. 2017. "Linoleic Acid: A Nutritional Quandary." Healthcare 5(2):25.

412. Adam, O., Wolfram, G., et al. 1986. "Effect of alpha-linolenic acid in the human diet on linoleic acid metabolism and prostaglandin biosynthesis." J Lipid Res 27(4):421-6.

413. Goyens, P.L., Spilker, M.E., et al. 2006. "Conversion of alpha-linolenic acid in humans is influenced by the absolute amounts of alpha-linolenic acid and linoleic acid in the diet and not by their ratio." Am J Clin Nutr 84(1):44-53.

414. Angela Liou, Y., and Innis, S.M. 2009. "Dietary linoleic acid has no effect on arachidonic acid, but increases n-6 eicosadienoic acid, and lowers dihomo-gamma-linolenic and eicosapentaenoic acid in plasma of adult men." Prostaglandins Leukot Essent Fatty Acids 80(4):201-6.

415. Simopoulos, A.P. 2006. "Evolutionary aspects of diet, the omega-6/omega-3 ratio and genetic variation: nutritional implications for chronic diseases." Biomed Pharmacother 60(9):502-7.

416. Simopoulos, A.P. 2002. "The importance of the ratio of omega-6/omega-3 essential fatty acids." Biomed Pharmacother 56(8):365-79.

417. Simopoulos, A.P. 2008. "The importance of the omega-6/omega-3 fatty acid ratio in cardiovascular disease and other chronic diseases." Exp Biol Med 233(6):674-88.

418. Simopoulos, A.P. 2010. "The omega-6/omega-3 fatty acid ratio: health implications." OCL 17(5):267–275.

419. Simopoulos, A.P. "The importance of the omega-6/omega-3 fatty acid ratio in cardiovascular disease and other chronic diseases."

420. The Cleveland Clinic. 2021. "Electrolytes." Accessed March 29, 2025. https://my.clevelandclinic.org/health/diagnostics/21790-electrolytes.

421. National Institutes of Health. 2022. "Potassium." Accessed March 29, 2025. https://ods.od.nih.gov/factsheets/Potassium-HealthProfessional/.

422. Goff, J.P. 2018. "Invited review: Mineral absorption mechanisms, mineral interactions that affect acid–base and antioxidant status, and diet considerations to improve mineral status." J of Dairy Sci 101(4): 2763-2813.

423. Felsenfeld, A.J., and Levine, B.S. 2015. "Pathophysiology of Calcium, Phosphorus, and Magnesium in Chronic Kidney Disease." Chronic Renal Disease Amsterdam, The Netherlands: Elsiver.

424. Areco, V.A., Kohan, R., et al. 2020. "Intestinal Ca2+ absorption revisited: A molecular and clinical approach." World J Gastroenterol 26(24):3344-3364.

425. The Cleveland Clinic. 2022. "Electrolyte Imbalance." Accessed March 29, 2025. https://my.clevelandclinic.org/health/symptoms/24019-electrolyte-imbalance.

426. National Academies of Sciences, Engineering, and Medicine. 1989. "Water and Electrolytes." In Recommended Dietary Allowances:

10th Edition Washington, DC: The National Academies Press.

427. Milton, K. 2003. "Micronutrient intakes of wild primates: are humans different?" Comp Biochem Physiol A Mol Integr Physiol 136(1):47-59.

428. Phillips, K.M., Pehrsson, P.R., et al. 2014. "Nutrient composition of selected traditional United States Northern Plains Native American plant foods." J Food Compos Anal 34(2):136-152.

429. Wolf, Robb. 2025. "Electrolytes and heart health: A science-based guide." Accessed March 29, 2025. https://science.drinklmnt.com/electrolytes/electrolytes-and-heart-health/.

430. Svetkey, L.P., McKeown, S.P., et al. 1996. "Heritability of salt sensitivity in black Americans." Hypertension 28(5):854-8.

431. He, J., Mills, K.T., et al. 2016. "Urinary Sodium and Potassium Excretion and CKD Progression." J Am Soc Nephrol 27(4):1202-12.

432. Willey, J., Gardener, H., et al. 2017. "Dietary Sodium to Potassium Ratio and Risk of Stroke in a Multiethnic Urban Population: The Northern Manhattan Study." Stroke 48(11):2979-2983.

433. Kim, B.S., Yu, M.Y., et al. 2024. "Effect of low sodium and high potassium diet on lowering blood pressure and cardiovascular events." Clin Hypertens 30(1):2.

434. American Heart Association. 2024. "How Potassium Can Help Prevent or Treat High Blood Pressure." Accessed March 29, 2025. https://www.heart.org/en/health-topics/high-blood-pressure/changes-you-can-make-to-manage-high-blood-pressure/how-potassium-can-help-control-high-blood-pressure.

435. Nieves, J.W. 2005. "Osteoporosis: the role of micronutrients." Am J Clin Nutr 81(5): 1232S-1239S.

436. Mahendra, A. 2023. "An overview on electrolytes: Its importance, function, and imbalances." Perspective 43(1).

437. Poles, J., Karhu, E., et al. 2021. "The effects of twenty-four nutrients and phytonutrients on immune system function and inflamma-

tion: A narrative review." J Clin Transl Res 7(3):333-376.

438. Gupta, C., and Prakash, D. 2014. "Phytonutrients as therapeutic agents." J Complement Integr Med 11(3):151-169.

439. Murillo, A.G., Hu, S., et al. 2019. "Zeaxanthin: Metabolism, Properties, and Antioxidant Protection of Eyes, Heart, Liver, and Skin." Antioxidants 8(9):390.

440. Khoo, H.E., Azlan, A., et al. 2017. "Anthocyanidins and anthocyanins: colored pigments as food, pharmaceutical ingredients, and the potential health benefits." Food Nutr Res 61(1):1361779.

441. Rahimi, P., Abedimanesh, S., et al. 2019. "Betalains, the nature-inspired pigments, in health and diseases." Crit Rev Food Sci Nutr 59(18):2949-2978.

442. Evans, J.A., and Johnson, E.J. 2010. "The role of phytonutrients in skin health." Nutrients 2(8):903-28.

443. Martins, T., Barros, A.N., et al. 2023. "Enhancing Health Benefits through Chlorophylls and Chlorophyll-Rich Agro-Food: A Comprehensive Review." Molecules 28(14):5344.

444. Rana, A., Samtiya, M., et al. 2022. "Health benefits of polyphenols: A concise review." J Food Biochem 46(10):e14264.

445. WebMD Editorial Contributors. 2024. "What to Know About Lactic Acid in Food." Accessed March 29, 2025. https://www.webmd.com/diet/what-to-know-about-lactic-acid-food.

446. Huang, W., Hu, W., et al. 2021. "Acetate supplementation produces antidepressant-like effect via enhanced histone acetylation." J Affect Disord 28:51-60.

447. Gunnars, K. 2024. "6 Health Benefits of Apple Cider Vinegar, Backed by Science." Accessed March 29, 2025. https://www.healthline.com/nutrition/6-proven-health-benefits-of-apple-cider-vinegar.

448. Binu, S. 2024. "Malic Acid: Uses, Health Benefits And Side Effects." Accessed March 29, 2025. https://www.netmeds.com/health-library/post/malic-acid-uses-health-benefits-and-side-effects.

449. Sautin, Y.Y., and Johnson, R.J. 2008. "Uric acid: the oxidant-antioxidant paradox." *Nucleos Nucleot Nucl* 27(6):608-19.

9 798995 680901